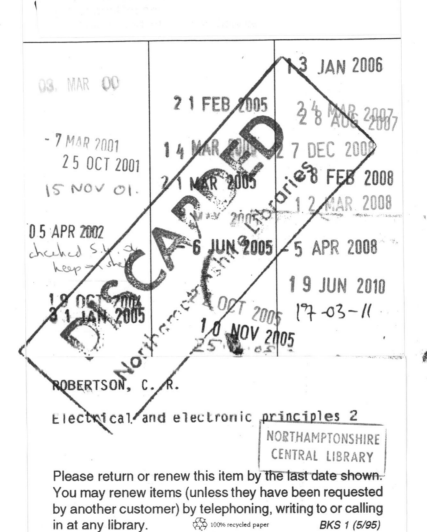
Please return or renew this item by the last date shown.
You may renew items (unless they have been requested
by another customer) by telephoning, writing to or calling
in at any library.   100% recycled paper   *BKS 1 (5/95)*

ELECTRICAL AND ELECTRONIC PRINCIPLES
■ 2 ■

Christopher R Robertson

Formerly Head of School
Technology Division
West Kent College
Tonbridge
Kent

Edward Arnold
A member of the Hodder Headline Group
LONDON SYDNEY AUCKLAND

Edward Arnold is a division of Hodder Headline PLC
338 Euston Road, London NW1 3BH

© 1993 Christopher R Robertson

First published in the United Kingdom 1993

2 4 6 7 5 3
94 96 98 97 95

British Library Cataloguing in Publication Data
Robertson, C R
Electrical and Electronic Principles 2
I. Title
621.3

ISBN 0-340-59231-1

Typeset in 10/11 Plantin by Wearset, Boldon, Tyne and Wear
Printed and bound in the United Kingdom by The Bath Press, Avon

▪ Contents ▪

Preface   v

Introduction   vi

1 Single-Phase a.c. Circuits
Summary of series circuits and equations—R-C parallel circuits—R-L parallel   1
circuits—R-L-C parallel circuits—Practical components in parallel—Series
resonance—Circuit Q-factor—Frequency response—Parallel resonance—Power
factor and power factor correction—Filters—Coupled tuned circuits—
Assignment questions—Suggested practical assignments.

2 Three-Phase a.c. Circuits
Generation of three-phase supply—Three-phase, 4-wire system—Relationship   28
between line and phase quantities in star—Delta or mesh connection—
Relationship between line and phase quantities in delta—Power dissipation—
Star/delta supplies and loads—Measurement of three-phase power—
Two-wattmeter method—Neutral current in unbalanced load—Advantages of
three-phase systems—Assignment questions—Suggested practical assignments.

3 Network Theorems and Attenuator Circuits
Superposition theorem—Constant voltage and constant current sources   48
Thévénin's theorem—Norton's theorem—The maximum power transfer
theorem—The decibel and its usage—Attenuator networks—Assignment
questions—Suggested practical assignments.

4 D.C. Transients
Capacitor-resistor series circuit (charging)—Capacitor-resistor series circuit   70
(discharging)—Inductor-resistor series circuit (connection to supply)—
Inductor-resistor series circuit (disconnection)—Affect of time constant on
alternating waveforms—Differentiating and integrating networks—Assignment
questions—Suggested practical assignments.

5 A.C. Machines
Transformer emf equation—The practical transformer—Measurement of   88
transformer losses—Transformer efficiency—Alternators—Rotor
construction—Alternator emf equation—Alternator losses—Production of a
rotating field from a polyphase supply—Three-phase induction motor—Three-
phase synchronous motor—Starting methods for three-phase motors—Single-
phase motors—Assignment questions—Suggested practical assignments.

6 D.C. Machines
Motor/generator duality—The generation of d.c. voltage—The commutation   112
process—Armature reaction—Construction of d.c. machines—Types of

armature winding—Generator emf equation—Classification of generators—
Separately excited generator—Shunt generator—Series generator—Compound
generator—D.C. motors—Shunt motor—Series motor—Compound motors—
Separately excited motor—The importance of back-emf—D.C. faceplate
starter—Efficiency of d.c. machines—Simple methods of speed control—
Stepper motors—Assignment questions—Suggested practical assignments.

7 Measuring Instruments and Measurements
Introduction (measurement errors)—Accuracy and sensitivity—Total   142
measurement error—Wattmeter corrections—Electronic voltmeter—Digital
voltmeter—Complex waveforms—Measurement of phase and frequency—A.C.
bridges

8 Control Principles
System elements and analogies—First-order systems—Ideal second-order   153
systems—Practical second-order systems—Driven damped systems—Control
system strategies—Computer control of systems—Assignment questions.

9 Information Transmission
Transmission systems—Digital signals—Serial and parallel transmission—   173
Asynchronous data transmission—Synchronous data transmission—
Comparison of asynchronous and synchronous transmission—
Protocols—High-level data link control—Data network configurations
(topologies)—Local area network (LAN)—Wide area network (WAN)—
Assignment questions.

10 Modulation Techniques
Amplitude modulation (am)—modulation index—Power in the am waveform—   184
Demodulation—Frequency modulation (fm)—Frequencies present in the fm
waveform—Power in the fm waveform—Comparison of amplitude and
frequency modulation—Phase modulation—Frequency-shift keying (fsk)—
Pulse modulation—Time-division multiplexing—Assignment questions.

Answers to Assignment Questions   202

Index   205

▪ **Preface** ▪

This second volume of two books completes the coverage of the BTEC Bank of Objectives for Electrical and Electronic Principles. The topics covered in this book are those considered to be at the NIII level. The Bank of Objectives is considered to contain material sufficient for three modules (units). This book contains material equivalent to approximately one-and-a-half modules, therefore some sections may not apply to particular programmes of study. The breadth of subject matter does however make it suitable for use by students other than those embarked on a BTEC programme of study.

C. Robertson
Tonbridge, Kent. March 1993

▪ Introduction ▪

Almost every chapter contains a number of worked examples both to illustrate the application of the theory, and to highlight certain principles. There are also Assignment Questions at the ends of the chapters, which are intended to give the student further practice in applying the knowledge gained. The vast majority of these questions require numerical solutions, and the answers to these are contained in a section at the end of the book.

The contents of the first four chapters, and Chapter 7, would normally be considered as material common to any BTEC programme at level NIII. The remaining chapters enable choices for the completion of a programme of study.

As with volume 1 of this title, where a word or phrase is shown in **bold type**, a box with a shaded border will be found close by. Within this box will be found a brief explanation of the words used in the text.

▪ 1 Single-Phase a.c. Circuits ▪

This chapter deals with the solution of parallel circuits, and the effects of both series and parallel resonance. Also included are the technique of power factor correction, and the concepts of tuned circuits and filters.

On completion of the chapter, you should be able to:

1] Determine the current flows, p.d.s and power dissipation in parallel a.c. circuit configurations.

2] State the conditions for resonance in both series and parallel circuits, and apply the formulae to obtain the resonant frequency in each case.

3] Define and calculate circuit Q-factor.

4] Appreciate the need for power factor correction, and carry out the relevant calculations.

5] Understand the effect of turned (resonant) circuits on circuit selectivity and bandwidth, and their application to filter circuits.

1.1 Summary of Series a.c. Circuits

The solution of parallel a.c. circuits is a simple extension of the methods used for series circuits. It is therefore sensible, at this stage, to summarise what has been learned about series circuits.

1 In a series circuit, the current is common to all of the series elements. For this reason the current is chosen as the reference phasor.

2 In a series circuit, the *phasor* sum of the p.d.s equals the total applied voltage.

3 Reactance is the opposition offered, to the flow of a.c., by a pure inductor or capacitor.

Reactance is measured in ohms, and the symbols are X_L and X_C.

4 Reactance depends upon the frequency of the supply, such that:

$$X_L = 2\pi f_L \text{ ohm}$$

$$\text{and } X_C = \frac{1}{2\pi fC} \text{ ohm}$$

5 Impedance is the overall opposition offered to the flow of a.c., in a circuit that contains both resistance and reactance. Impedance has the symbol Z, and is measured in ohms.

6
$$Z = \frac{V}{I} = \sqrt{R^2 + (X_L - X_C)^2} \text{ ohm}$$

7 Power is dissipated *only* by a resistive component. The power may be calculated from either:

$$P = I^2 R \quad \text{or} \quad P = VI \cos \phi \text{ watt}$$

where ϕ is the circuit phase angle, and $\cos \phi$ is the circuit power factor.

8 For a pure resistor,

$$\phi = 0° \qquad V_R \text{ in phase with } I$$

For a pure inductor,

$$\phi = +90° \quad V_L \text{ leads } I \text{ by } 90°$$

For a pure capacitor,

$$\phi = -90° \quad V_C \text{ lags } I \text{ by } 90°$$

the above being summed up by the 'keyword' CIVIL. Typical circuit and phasor diagrams are shown in Figs. 1.1 and 1.2.

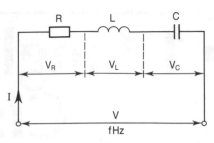

Fig 1.1

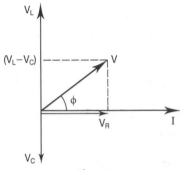

Fig 1.2

1.2 The *R-C* Parallel Circuit

Consider a pure resistor and a pure capacitor, connected in parallel across an a.c. supply, as shown in Fig. 1.3. Since both components are

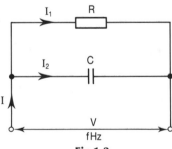

Fig 1.3

connected directly to the same supply terminals, then the *supply voltage*, V, is common to them both. On the other hand, each component will draw a current dependent upon its opposition offered.

i.e. $\qquad I_1 = \dfrac{V}{R} \text{ amp, and } I_2 = \dfrac{V}{X_C} \text{ amp}$

However, both I_1 and I_2 are supplied from the a.c. source. Thus the total current drawn from the supply, I is the PHASOR SUM of I_1 and I_2. The resulting phasor diagram, as shown in Fig. 1.4,

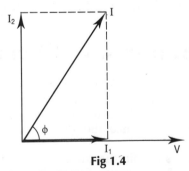

Fig 1.4

therefore uses the applied voltage V as the reference phasor. From this diagram, it may be seen that

$$I = \sqrt{I_1^2 + I_2^2} \text{ amp}$$

and $\cos \phi = \dfrac{I_1}{I} \qquad \text{or} \tan \phi = \dfrac{I_2}{I_1}$

Note: In a parallel circuit, the equation $Z = \sqrt{R^2 + X_C^2}$ CAN NOT BE USED to obtain the circuit impedance. The impedance can only be obtained from $Z = V/I$ ohm. Also note the fact that since the current through the capacitor (I_2) *leads* the voltage V, then a capacitor in a parallel circuit results in a leading phase angle, ϕ. This may be confirmed by considering the word CIVIL.

▪ Worked Example 1.1 ▪

 A 2.2 μF capacitor is connected in parallel with a 1.5 kΩ resistor across a 100 V, 50 Hz supply. Sketch the circuit and phasor diagrams, and calculate (a) the current flow through each component, (b) the current drawn from the supply, (c) the circuit phase angle, and (d) the power dissipated.

A ————————————————————

$C = 2.2 \times 10^{-6}$ F; $R = 1500$ Ω; $V = 100$ V; $f = 50$ Hz

(a) $I_1 = \dfrac{V}{R}$ amp $= \dfrac{100}{1500}$

so $I_1 = 66.7$ mA **Ans**

$X_C = \dfrac{1}{2\pi f C}$ ohm $= \dfrac{1}{2 \times \pi \times 50 \times 2.2 \times 10^{-6}}$

so $X_C = 1446.9$ Ω

$I_2 = \dfrac{V}{X_C}$ amp $= \dfrac{100}{1446.9}$

so $I_2 = 69.1$ mA **Ans**

(b) $I = \sqrt{I_1^2 + I_2^2}$ amp $= \sqrt{66.7^2 + 69.1^2}$ mA

hence $I = 96$ mA **Ans**

(c) $\phi = \cos^{-1}\dfrac{I_1}{I} = \cos^{-1}\dfrac{66.7}{96}$

so $\phi = 46.03°$ leading **Ans**

(d) $P = VI \cos\phi$ watt $= 100 \times 96 \times 10^{-3} \times 0.6943$

so $P = 6.67$ W **Ans**

Alternatively, $P = I_1^2 R$ watt $= (66.7 \times 10^{-3})^2 \times 1500$

so $P = 6.67$ W **Ans**

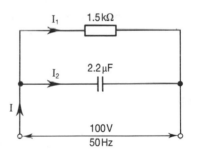

Fig 1.5

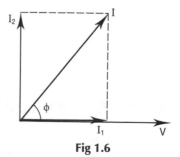

Fig 1.6

▪ Worked Example 1.2 ▪

Q For the circuit shown in Fig. 1.7, determine (a) the supply voltage, (b) the supply frequency, and (c) the current drawn from the supply.

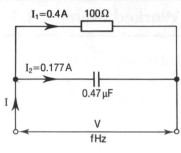

Fig 1.7

A ───

(a) Since the supply is connected directly across the resistor, then the p.d. across $R = V$.

therefore, $V = I_1R$ volt $= 0.4 \times 100$

so, $V = 40$ V **Ans**

(b) $$X_C = \frac{V}{I_2} \text{ ohm} = \frac{40}{0.177}$$

therefore, $X_C = 225.99$ Ω

and, since $X_C = \dfrac{1}{2\pi fC}$ ohm

then, $f = \dfrac{1}{2\pi C X_C}$ hertz

so, $f = \dfrac{1}{2 \times \pi \times 225.99 \times 0.47 \times 10^{-6}}$

hence, $f = 1498$ Hz **Ans**

(c) $I = \sqrt{I_1^2 + I_2^2}$ amp $= \sqrt{0.4^2 + 0.177^2}$

thus, $I = 0.437$ A **Ans**

1.3 The *R-L* Parallel Circuit ─────────────────────

When a pure inductor is connected in parallel with a resistor, the techniques used to analyse the circuit are the same as for the *C-R* circuit. The only difference will be that the circuit will have a lagging phase angle. The following example illustrates this.

▪ Worked Example 1.3 ▪

Q The circuit of Fig. 1.8 shows a pure inductor connected in parallel with a 200 Ω resistor, across a 50 V, 1 kHz supply. If the current drawn from the supply is 1.03 A at a power factor of 0.2427, determine (a) the current in each branch, (b) the power dissipated, and (c) the value of the inductance.

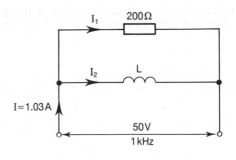

Fig 1.8

A

$V = 50$ V; $f = 10^3$ Hz; $I = 1.03$ A; $\cos \phi = 0.2427$

The corresponding phasor diagram is shown in Fig. 1.9.

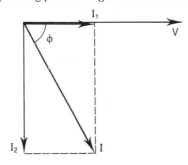

Fig 1.9

(a) $\phi = \cos^{-1} 0.2427 = 75.95°$

$$I_1 = \frac{V}{R} \text{ amp} = \frac{50}{200}$$

hence, $I_1 = 0.25$ A **Ans**

From the phasor diagram, it may be seen that:

$$I_2 = \sqrt{I^2 - I_1^2} \text{ amp} = \sqrt{1.03^2 - 0.25^2}$$

therefore, $I_2 = 1$ A **Ans**

(b) $P = VI \cos \phi \text{ watt} = 50 \times 1.03 \times 0.2427$

so $P = 12.5$ W **Ans**

(c) $$X_L = \frac{V}{I_2} \text{ ohm} = \frac{50}{1}$$

so, $X_L = 50$ Ω

but, $X_L = 2\pi fL$ ohm,

so, $L = \frac{X_L}{2\pi f} \text{ henry} = \frac{50}{2 \times \pi \times 10^3}$

hence, $L = 7.96$ mH **Ans**

1.4 *R-L-C* Parallel Circuit

This circuit is a simple extension of the previous two considered. The only differences are that, (a) there are three branch currents, and (b) the circuit phase angle may be either leading or lagging. The latter depends upon the relative values of the currents through inductor and capacitor.

■ Worked Example 1.4 ■

A 75 Ω resistor, an 80 mH inductor, and a 40 μF capacitor are connected in parallel across an 80 V, 100 Hz supply. Sketch the circuit and phasor diagrams, and calculate (a) the three branch currents, (b) the current drawn from the supply, and (c) the circuit phase angle and power factor.

A

$R = 75$ Ω; $L = 80 \times 10^{-3}$ H; $C = 44 \times 10^{-6}$ F; $V = 80$ V;
$f = 100$ Hz

The circuit and phasor diagrams are shown in Figs. 1.10 and 1.11 respectively.

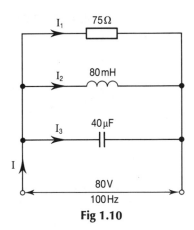

Fig 1.10

(a) $I_1 = \dfrac{V}{R}$ amp $= \dfrac{80}{75}$

so, $I_1 = 1.067$ A **Ans**

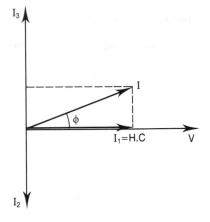

Fig 1.11

$X_L = 2\pi f L$ ohm $= 2 \times \pi \times 100 \times 80 \times 10^{-3}$

$X_L = 50.265$ Ω

$I_2 = \dfrac{V}{X_L}$ amp $= \dfrac{80}{50.265}$

and, $I_2 = 1.592$ A **Ans**

$X_C = \dfrac{1}{2\pi f C}$ ohm $= \dfrac{1}{2 \times \pi \times 100 \times 40 \times 10^{-6}}$

so $X_C = 39.8 \ \Omega$

$$I_3 = \frac{V}{X_C} \ \text{amp} = \frac{80}{39.8}$$

hence, $I_3 = 2.01$ A **Ans**

(b) From the phasor diagram:

Horizontal component, HC $= I_1 = 1.067$ A

Vertical component, VC $= I_3 - I_2 = 2.01 - 1.592$

so, VC $= 0.418$ A

$$I = \sqrt{VC^2 + HC^2} = \sqrt{0.418^2 + 1.067^2}$$

therefore, $I = 1.146$ A **Ans**

$$\phi = \tan^{-1} \frac{VC}{HC} = \tan^{-1} \frac{0.418}{1.067}$$

hence, $\phi = 21.4°$ leading **Ans**

power factor, $\cos \phi = 0.931$ leading **Ans**

1.5 Practical Components in Parallel

Thus far, we have considered that the circuit components are perfect: i.e. pure resistor, pure inductor and pure capacitor. In practice, unless the resistor is of the wirewound type, it is normally assumed to be purely resistive. Similarly, provided that the frequency is not very high, or the capacitor is not an electrolytic type, then the capacitor may also be considered as perfect. The inductor, on the other hand, may rarely be considered as pure inductance. Thus, the resistance of the inductor should be taken into account. The inductor coil therefore has an impedance, Z_{coil} ohm.

A circuit commonly met in practice consists of a 'practical' inductor, connected in parallel with a 'perfect' capacitor. Such a circuit is illustrated in

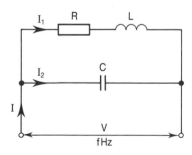

Fig 1.12

Fig. 1.12, which could represent the coil of an a.c. motor, with a parallel-connected capacitor.

The current through the coil, $I_1 = \dfrac{V}{Z_1}$ amp

where impedance of coil $= Z_1 = \sqrt{R^2 + X_L^2}$ ohm

and $\phi_1 = \cos^{-1} \dfrac{R}{Z_1}$ (lagging V)

Current through the capacitor, $I_2 = V/X_C$ amp and this current will lead the applied voltage by 90°.

The phasor diagram is shown in Fig. 1.13. From

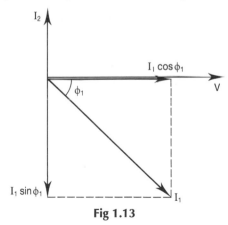

Fig 1.13

this diagram it may be seen that the capacitor current forms a vertical component only. The inductor coil current has both horizontal and vertical components, $I_1 \cos \phi_1$ and $I_1 \sin \phi_1$ respectively. The total vertical component (VC) is therefore the difference between I_2 and $I_1 \sin \phi_1$. A supplementary phasor diagram is shown in Fig. 1.14, from which the circuit current and phase angle can be determined.

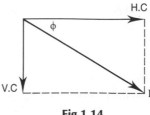

Fig 1.14

■ **Worked Example 1.5** ■

 A coil of inductance 159.2 mH and resistance 25 Ω is connected in parallel with a 40 μF capacitor, across a 240 V, 50 Hz supply. Sketch the circuit and phasor diagrams, and calculate (a) the current through, and phase angle of, the coil, (b) the capacitor current, (c) the supply current and phase angle, (d) the power consumed and (e) the circuit impedance.

A

$L = 159.2 \times 10^{-3}$ H; $R = 25$ Ω; $C = 40 \times 10^{-6}$ F; $V = 240$ V; $f = 50$ Hz

The circuit and phasor diagrams are shown in Figs. 1.15 and 1.16 respectively.

(a) $\qquad X_L = 2\pi f L$ ohm $= 2 \times \pi \times 50 \times 159.2 \times 10^{-3}$

so, $X_L = 50$ Ω

$\qquad Z_1 = \sqrt{R^2 + X_L^2}$ ohm $= \sqrt{25^2 + 50^2}$

and $Z_1 = 55.9$ Ω

$$I_1 = \frac{V}{Z_1} \text{ amp} = \frac{240}{55.9}$$

hence, $I_1 = 4.29$ A **Ans**

$$\phi_1 = \cos^{-1} \frac{R}{Z_1} = \cos^{-1} \frac{25}{55.9}$$

therefore $\phi_1 = -63.43°$ **Ans**

(b) $\qquad X_C = \frac{1}{2\pi f C}$ ohm $= \frac{1}{2 \times \pi \times 50 \times 40 \times 10^{-6}}$

so, $X_C = 79.58$ Ω

$$I_2 = \frac{V}{X_C} \text{ amp} = \frac{240}{79.58}$$

hence, $I_2 = 3.02$ A **Ans** (I_2 leads V by 90°)

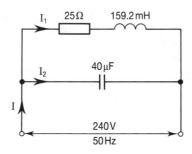

Fig 1.15

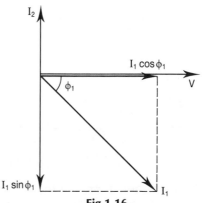

Fig 1.16

(c) Referring to Fig. 1.17:

$$HC = I_1 \cos \phi_1 \text{ amp} = 4.29 \cos 63.43°$$

so, $HC = 1.919 \text{ A}$

$$VC = I_2 - I_1 \sin \phi_1 \text{ amp} = 3.02 - 4.29 \sin 63.43°$$

and $VC = -0.817 \text{ A}$

$$I = \sqrt{VC^2 + HC^2} \text{ amp} = \sqrt{-0.817^2 + 1.919^2}$$

therefore, $I = 2.09 \text{ A}$ **Ans**

$$\phi = \tan^{-1}\frac{VC}{HC} = \tan^{-1}\frac{-0.817}{1.919}$$

hence, $\phi = -23.06°$ **Ans**

(d) $P = VI \cos \phi \text{ watt} = 240 \times 2.09 \times 0.9201$

so, $P = 461.5 \text{ W}$ **Ans**

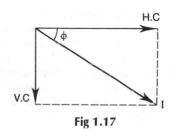

Fig 1.17

Alternatively, $P = I_1^2 R \text{ watt}$

(e)
$$Z = \frac{V}{I} \text{ ohm} = \frac{240}{2.09}$$

therefore, $Z = 114.8 \, \Omega$ **Ans**

Note: In both of the last two examples, the circuit current is *less* than both the current through the coil and that through the capacitor. This would be an impossible situation in the corresponding d.c. circuit, where $I = I_1 + I_2$ amp. It is possible in the parallel a.c. circuit because I is the PHASOR SUM of the branch currents, and NOT the arithmetic sum.

1.6 Series Resonance

The concept of resonance in a series R–L–C circuit was briefly dealt with in Volume 1. It will be helpful to review this concept now, and then to study the effect in more detail.

1 The reactances of both a capacitor and an inductor depend on the frequency of the supply.

2 $X_L \propto f$, and $X_C \propto 1/f$.

3 For given values of L and C; at low frequencies X_L will be relatively small, compared to X_C. At high frequencies, X_L will be relatively large, compared to X_C.

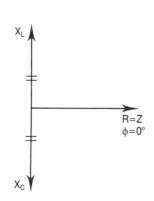

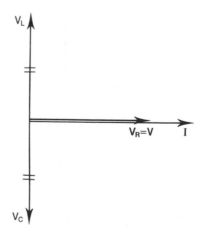

Fig 1.18

4 At one particular frequency, $X_L = X_C$. This is known as the resonant frequency, f_0 hertz.

5 Under resonant conditions, the circuit current and voltage are in phase with each other. The circuit therefore behaves as if it were purely resistive in nature. This is illustrated in the phasor diagram of Fig. 1.18.

6 The circuit impedance Z is at its minimum possible value of R ohm.

7 The resonance frequency, $f_0 = \dfrac{1}{2\pi\sqrt{LC}}$ hertz

or $\omega_0 = \dfrac{1}{\sqrt{LC}}$ rad/s

■ Worked Example 1.6 ■

 Q A 47 pF capacitor is connected in series with a coil of resistance 25 Ω and inductance 50 mH, across a 5 mV variable frequency supply. Calculate (a) the supply frequency that results in resonance, (b) the circuit current, (c) the p.d. across the capacitor and coil.

A

$C = 47 \times 10^{-12}\text{F}; R = 25\ \Omega; L = 50 \times 10^{-3}\ \text{H};$
$V = 5 \times 10^{-3}\ \text{V}$

The circuit and phasor diagrams are shown in Figs. 1.19 and 1.20 respectively.

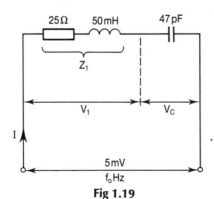

Fig 1.19

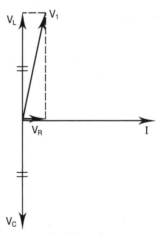

Fig 1.20

(a)

$f_0 = \dfrac{1}{2\pi\sqrt{LC}}\text{hertz} = \dfrac{1}{2\pi\sqrt{50 \times 10^{-3} \times 47 \times 10^{-12}}}$

therefore, $f_0 = 103.82\ \text{kHz}$ **Ans**

(b) At f_0, $Z = R$ ohm $= 25\ \Omega$

and $I = \dfrac{V}{Z}$ amp $= \dfrac{5 \times 10^{-3}}{25}$

hence, $I = 200\ \mu\text{A}$ **Ans**

(c) $\quad X_C = \dfrac{1}{2\pi fC}$ ohm $= \dfrac{1}{2\pi \times 103.82 \times 10^3 \times 47 \times 10^{-12}}$

so, $X_C = 32.62$ kΩ

$V_C = IX_C$ volt $= 200 \times 10^{-6} \times 32.62 \times 10^3$

therefore, $V_C = 6.52$ V **Ans**

Since the coil possesses both inductance and resistance,

then the total opposition it offers to the current is its impedance, $Z_1 = \sqrt{R^2 + X_L^2}$ ohm.
At resonance, $X_L = X_C = 32.62$ kΩ

so, $Z_1 = \sqrt{25^2 + (32.62 \times 10^3)^2} = 32.62$ kΩ

i.e. the coil resistance happens to be negligible compared with its reactance. Thus, the p.d. across the coil is the same as that across the capacitor, namely 6.52 V **Ans**

1.7 Circuit *Q*-factor

In the previous example, it may be seen that the p.d.s across the capacitor and the inductor are not only equal, but are many times greater than the supply voltage. In this particular case, $V_C = 1305 \times V$.

This effect is known as voltage magnification. The ratio of the p.d. across the capacitor (or inductor) to the applied voltage is called the voltage magnification factor, or *Q*-factor.

$$\text{Therefore, } Q = \frac{V_C}{V} = \frac{V_L}{V} \qquad (1.1)$$

but, $V_C = IX_C$, $V_L = IX_L$, and $V = IR \quad (Z = R)$

$$\text{therefore, } Q = \frac{IX_C}{IR} = \frac{IX_L}{IR}$$

$$\text{so, } Q = \frac{X_C}{R} = \frac{X_L}{R} \qquad (1.2)$$

also, since $X_C = 1/\omega_0 C$, and $X_L = \omega_0 L$ ohm

$$\text{then, } Q = \frac{1}{\omega_0 CR} = \frac{\omega_0 L}{R} \qquad (1.3)$$

Both equations (1.2) and (1.3) above require the resonant frequency to be known. However, the *Q*-factor may also be expressed in terms of the circuit components only, as follows:

$$Q = \frac{1}{\omega_0 CR}, \quad \text{and} \quad \omega_0 = \frac{1}{\sqrt{LC}}$$

$$\text{therefore, } Q = \frac{\sqrt{LC}}{CR}$$

$$\text{hence, } Q = \frac{1}{R}\sqrt{\frac{L}{C}} \qquad (1.4)$$

▪ Worked Example 1.7 ▪

 A series circuit consists of a 50 Ω resistor, a 0.15 H inductor and a 22 nF capacitor. This combination is connected across a 24 V a.c. supply. Determine (a) the resonant frequency, (b) the *Q*-factor, and (c) the p.d. across the capacitor and inductor.

A ─────────────────────────────────

$R = 50$ Ω; $L = 0.15$ H; $C = 22 \times 10^{-9}$ F; $V = 24$ V

(a) $\quad \omega_0 = \dfrac{1}{\sqrt{LC}}$ rad/s $= \dfrac{1}{\sqrt{0.15 \times 22 \times 10^{-9}}}$

so, $\omega_0 = 17.408 \times 10^3$ rad/s

hence, $f_0 = \dfrac{17.408 \times 10^3}{2\pi} = 2.77$ kHz **Ans**

(b)
$$Q = \frac{1}{R}\sqrt{\frac{L}{C}} = \frac{1}{50}\sqrt{\frac{0.15}{22 \times 10^{-9}}}$$

so, $Q = 52.22$ **Ans**

(c) $V_C = Q \times V$ volt $= 52.22 \times 24$

hence, $V_C = V_L = 1.253$ kV **Ans**

1.8 Frequency Response Curve

We have seen that, at the resonant frequency, the impedance of the series circuit reaches its minimum value of R ohm. At frequencies greater than and less than f_0, there is an overall reactive component, which results in increased impedance. Thus the current flow through the circuit reaches its maximum possible value at the resonant frequency. A graph of the circuit current versus frequency is known as the frequency response curve, and will have the shape shown in Fig. 1.21.

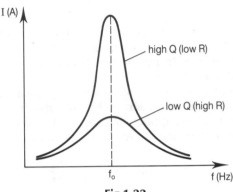

Fig 1.22

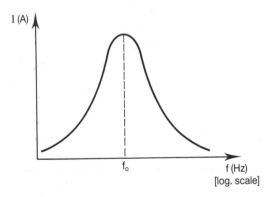

Fig 1.21

Note that the frequency axis is often plotted to a logarithmic scale (as in the diagram). The main reason for this is to enable a large range of frequencies to be accommodated, without the need for excessively wide graph paper. It also has the effect of making the graph more symmetrical about the vertical axis, on either side of f_0.

The Q-factor of the circuit will determine the 'peakiness' of the response curve. A high value of Q results in a curve with a high peak and steep sides on either side of f_0. A low value of Q results in a shallow curve, with gently sloping sides. From equations (1.2) to (1.4), for Q-factor, it is apparent

that a low value of R gives a high Q, and vice-versa. This effect is illustrated in Fig. 1.22.

From the curves shown in this diagram, it is apprent that a circuit having a high Q-factor will offer minimal opposition to only a narrow range of frequencies on either side of f_0. This property of the circuit is referred to as its selectivity. The range of frequencies concerned is called the bandwidth of the circuit. In order to make comparisons of selectivity between circuits, the term bandwidth needs to be explained.

Let the current that flows at resonance be referred to as I_0. The bandwidth, B, of the circuit is identified as that range of frequencies over which the circuit current is greater than or equal to $I_0/\sqrt{2}$ ($I \geqslant I_0/\sqrt{2}$). The current value at the two extremes defined above are shown as I_1 and I_2 in Fig. 1.23. The 'cut-off' frequencies are shown as f_1 and f_2. The circuit bandwidth is therefore the difference between f_2 and f_1.

i.e. Bandwidth, $B = (f_2 - f_1)$ hertz   (1.5)

and $I_1 = I_2 = \dfrac{I_0}{\sqrt{2}}$ (1.6)

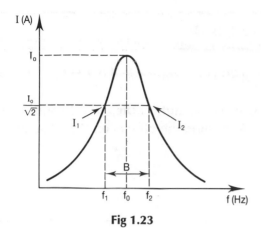

Fig 1.23

From Fig. 1.22, it should be obvious that the bandwidth for a circuit having a high Q-factor will be less than that for one with a low Q-factor. From this we can conclude that the bandwidth and Q-factor of a circuit are related to each other. This relationship is as follows:

$$\text{Bandwidth, } B = \frac{f_0}{Q} \text{ hertz} \qquad (1.7)$$

Thus far, we have seen that the Q-factor, and hence the selectivity of the circuit, is inversely proportional to the circuit resistance. Consider now equation (1.4), which is repeated below.

$$Q = \frac{1}{R}\sqrt{\frac{L}{C}}$$

From this equation it may be seen that Q is directly proportional to the ratio L/C. Thus, if the resistance is maintained constant and the ratio of L to C is varied, the Q-factor and selectivity will vary in direct proportion. This is illustrated in Fig. 1.24.

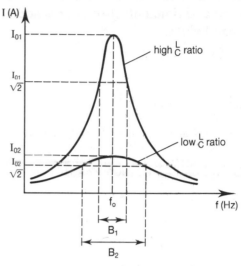

Fig 1.24

In a practical circuit it is difficult to vary the ratio of L to C without also varying R. If the resistance is kept constant by leaving the coil unchanged, then only C can be varied. However, when this is done, the circuit resonant frequency will also be varied.

■ Worked Example 1.8 ■

Q A tuned (resonant) circuit consists of a coil, having 40 Ω resistance and 0.2 H inductance, connected in series with a 22 nF capacitor.
(i) Determine (a) the circuit Q-factor, (b) the resonant frequency, and (c) the circuit bandwidth.
(ii) If an additional 40 Ω resistor is added to the circuit, determine the effects on the Q-factor and bandwidth.

A _____

(i) $R = 40\ \Omega;\ L = 0.2\ \text{H};\ C = 22 \times 10^{-9}\text{F}$

(a) $Q = \dfrac{1}{R}\sqrt{\dfrac{L}{C}} = \dfrac{1}{40}\sqrt{\dfrac{0.2}{22 \times 10^{-9}}}$

   therefore, $Q = 75.38$ **Ans**

(b) $f_0 = \dfrac{1}{2\pi\sqrt{LC}}\ \text{hertz} = \dfrac{1}{2\pi\sqrt{0.2 \times 22 \times 10^{-9}}}$

   so, $f_0 = 2.4\ \text{kHz}$ **Ans**

(c) $B = \dfrac{f_0}{Q}\ \text{hertz} = \dfrac{2.4 \times 10^3}{75.38}$

   hence, $B = 31.83\ \text{Hz}$ **Ans**

(ii) $R = 80\ \Omega;\ L$ and C as before

   Since R is now twice its original value, then Q must be halved, and B must be doubled. The resonant frequency will remain unchanged at 2.4 kHz.

   Thus, $Q = 37.69$ and $B = 63.76\ \text{Hz}$ **Ans**

A series circuit, at or near to its resonant condition, offers relatively little opposition to the flow of current through it. It is therefore also known as an acceptor circuit. This effect is utilised in a filter circuit, known as a bandpass filter. Filter circuits are discussed later in this chapter.

1.9 Parallel Resonance _____

Consider a coil possessing both inductance and resistance, connected in parallel with a capacitor, across an a.c. supply. This circuit is shown in Fig. 1.25. Let the frequency of the supply be the resonant frequency for the circuit. This will mean that the circuit current, I, will be in phase with the supply voltage, V. This condition is illustrated in Fig. 1.26. From this phasor diagram it may be seen that the resonant condition occurs when the two vertical components of current are equal. Thus resonance occurs when $I_1 \sin \phi_1 = I_2$. The result is that the current drawn from the supply, I is equal to $I_1 \cos \phi_1$.

Now, $I_1 = \dfrac{V}{Z_1}$, and considering the

impedance triangle for the coil (Fig. 1.27)

$$\sin \phi_1 = \dfrac{X_L}{Z_1} = \dfrac{\omega_0 L}{Z_1}$$

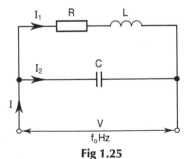

Fig 1.25

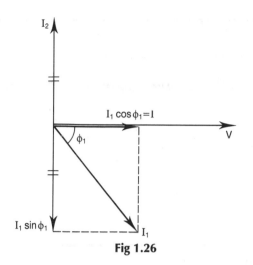

Fig 1.26

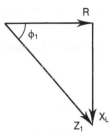

Fig 1.27

therefore, $I_1 \sin \phi_1 = \dfrac{V\omega_0 L}{Z_1^{\,2}}$ .... [1]

also, $I_2 = \dfrac{V}{X_C} = V\omega_0 C$ .... [2]

For resonance, [1] = [2], so

$\dfrac{V\omega_0 L}{Z_1^{\,2}} = V\omega_0 C$, and dividing through by $V\omega_0$

$\dfrac{L}{Z_1^{\,2}} = C$

so $Z_1^{\,2} = \dfrac{L}{C}$ .... [3]

but, $Z_1 = \sqrt{R^2 + X_L^{\,2}}$, so $Z_1^{\,2} = R^2 + X_L^{\,2}$

or, $Z_1^{\,2} = R^2 + (\omega_0 L)^2$,

and substituting this into equation [3]:

$R^2 + \omega_0^{\,2} L^2 = \dfrac{L}{C}$

$\omega_0^{\,2} L^2 = \dfrac{L}{C} - R^2$

$\omega_0^{\,2} = \dfrac{L}{L^2 C} - \dfrac{R^2}{L^2} = \dfrac{1}{LC} - \dfrac{R^2}{L^2}$

therefore, $\omega_0 = \sqrt{\dfrac{1}{LC} - \dfrac{R^2}{L^2}}$ rad/s (1.8)

and, $f_0 = \dfrac{1}{2\pi}\sqrt{\dfrac{1}{LC} - \dfrac{R^2}{L^2}}$ hertz (1.9)

Comparing equation (1.9) with that for series resonance, it will be seen that they are very similar. The difference is the inclusion of the term R^2/L^2 in equation (1.9). The significance of this is that, for the same circuit components, the resonant frequency for the parallel circuit is less than that for the series arrangement. However, if the term R^2/L^2 is very much less than $1/LC$, then the equation for parallel resonance *approximates* to:

$$f_0 \simeq \dfrac{1}{2\pi\sqrt{LC}} \text{ hertz}$$

This condition tends to occur when the resistance of the coil is very small compared with its reactance.

It should be noted that if the supply frequency is either reduced below, or increased above f_0, then the vertical components of current will no longer 'cancel out'. At frequencies other than f_0, the circuit current will therefore be greater. Hence, we can draw the conclusion that at resonance, the impedance of a parallel circuit reaches its MAXIMUM value. This is exactly opposite to the case for a series resonant circuit.

The impedance of the parallel circuit at resonance is referred to as its dynamic impedance, Z_D, or its dynamic resistance, R_D ohm. Since the circuit phase angle at resonance is zero, the circuit again behaves as if it were purely resistive in nature. However, Z_D is NOT EQUAL to the resistance of the coil, R, since this would be the *minimum* possible value. We therefore need to be able to calculate the value of Z_D for a given circuit.

From the phasor diagram (Fig. 1.26), and the impedance triangle (Fig. 1.27), we have:

$$\tan \phi_1 = \frac{\omega_0 L}{R} = \frac{I_1 \sin \phi_1}{I}$$

but, $I = \dfrac{V}{Z_D}$

so $\dfrac{\omega_0 L}{R} = \dfrac{I_1 \sin \phi_1 Z_D}{V}$

also, $\qquad I_1 \sin \phi_1 = I_2$

therefore, $\dfrac{\omega_0 L}{R} = \dfrac{I_2}{V} Z_D;$ and $\dfrac{I_2}{V} = \omega_0 C$

so $\dfrac{\omega_0 L}{R} = \omega_0 C Z_D;$ and dividing through by ω_0:

$$\frac{L}{R} = C Z_D$$

hence, $\quad Z_D = \dfrac{L}{CR}$ ohm $\qquad$ (1.10)

■ Worked Example 1.9 ■

Q A coil of inductance 15 mH and resistance 50 Ω, is connected in parallel with a 1 μF capacitor, across a 20 V variable frequency a.c. supply. Determine (a) the approximate and actual values for the resonant frequency, (b) the dynamic impedance, (c) the current drawn from the supply, under resonant conditions, and (d) the corresponding capacitor current.

A

$L = 15 \times 10^{-3}$ H; $R = 50\ \Omega$; $C = 10^{-6}$ F; $V = 20$ V

(a) $\quad f_0 = \dfrac{1}{2\pi\sqrt{LC}}$ hertz $= \dfrac{1}{2\pi\sqrt{15 \times 10^{-3} \times 10^{-6}}}$

so $f_0 = 1.3$ kHz **Ans** (approximate value)

$f_0 = \dfrac{1}{2\pi}\sqrt{\dfrac{1}{LC} - \dfrac{R^2}{L^2}}$ hertz

$= \dfrac{1}{2\pi}\sqrt{\dfrac{1}{15 \times 10^{-3} \times 10^{-6}} - \dfrac{50^2}{(15 \times 10^{-3})^2}}$

$= \dfrac{1}{2\pi}\sqrt{6.67 \times 10^7 - 1.11 \times 10^7}$

so, $f_0 = 1.19$ kHz **Ans** (actual value)

(b) $\quad Z_D = \dfrac{L}{CR}$ ohm $= \dfrac{15 \times 10^{-3}}{50 \times 10^{-6}}$

$Z_D = 300\ \Omega$ **Ans**

(c) $\quad I = \dfrac{V}{Z_D} = \dfrac{20}{300}$

hence, $I = 66.7$ mA **Ans**

(d) $\quad I_C = \dfrac{V}{X_C}$ amp; where $X_C = \dfrac{1}{2\pi f_0 C}$ ohm

so $I_C = V \times 2\pi f_0 C = 20 \times 2\pi \times 1.19 \times 10^3 \times 10^{-6}$

and, $I_C = 149.5$ mA **Ans**

1.10 The Importance of Power Factor

In situations where large amounts of power are generated and consumed, the power factor of the load being supplied becomes increasingly important. It is possible for the load, and its power factor, to change. For this reason, most a.c. electrical machines, such as alternators (generators) and transformers are rated in VA or kVA. The kVA rating of such a machine indicates the highest current that it can safely supply, at its rated output voltage. For example, a 350 V alternator rated at 525 kVA can supply a maximum possible current of 1500 A. The *power* output of the alternator would be 525 kW *only* if the load was purely resistive i.e. if the power factor was unity (cos φ = cos 0° = 1). For any power factor less than 1, the power output would be less than the kVA figure. This is of course the normal situation in practice.

Consider now, the above specified alternator, supplying a load whose power factor can be switched between (say) 1 and 0.5. The turbine driving the alternator must develop 525 kW plus the power losses in the alternator, in order to yield 525 kVA at the alternator output. If the load power factor is unity, then the load will develop a power of 525 kW, at a current of 1500 A. If the load power factor is 0.5, then it will dissipate only 262.5 kW, at a current of 1500 A, since $P = VI \cos \phi$ watt. The alternator however, still have to provide the same voltage and current as before. The load is therefore not making efficient use of the available power from the supply. This situation may be summed up by saying that, *for a given power*, the lower the load power factor, the larger must be the alternator to generate that power. The result is inefficient usage of the supply, and an increase of costs. For this reason the supply authorities always try to attempt to improve the power factor of their loads. They also encourage large industrial consumers to do likewise, by an additional tariff imposed on the reactive component (kVAr) supplied.

1.11 Power Factor Correction

We have seen that when a capacitor is connected in parallel with an inductive circuit unity power factor is obtained, at the resonant frequency. Alternatively, the resonant condition can be achieved by altering the value of the capacitor, so that the resonant and supply frequencies are one and the same. The vast majority of practical loads are inductive in nature, resulting in a power factor of less than 1. In these cases, the power factor may be improved (increased) by a parallel connected capacitor. Consider the following example.

▪ Worked Example 1.10 ▪

 Q A motor draws a current of 3.5 A at a lagging power factor of 0.6, when connected to a 240 V, 50 Hz supply. Calculate (a) the value of parallel connected capacitor that will correct this power factor to unity, and (b) the value of current now drawn from the supply.

A _____

$I_1 = 3.5$ A; $\cos \phi_1 = 0.6$; $V = 240$ V; $f = 50$ Hz

(a) The appropriate circuit and phasor diagrams are shown in Figs. 1.28 and 1.29 respectively.

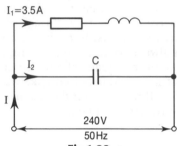

Fig 1.28

Thus, $I_2 = I_1 \sin \phi_1$ (see Fig. 1.29)

hence $I_2 = 3.5 \times 0.8 = 2.8$ A

$$X_C = \frac{V}{I_2} \text{ ohm} = \frac{240}{2.8}$$

therefore, $X_C = 85.71 \ \Omega$

$$X_C = \frac{1}{2\pi fC}; \text{ so } C = \frac{1}{2\pi fX_C} \text{ farad}$$

hence $C = \dfrac{1}{2 \times \pi \times 50 \times 85.71}$

therefore, $C = 37.14 \ \mu$F **Ans**

$\cos \phi_1 = 0.6$; so $\phi_1 = 53.13°$ and
$\sin \phi_1 = 0.8$

For unity power factor, the supply current must be in phase with the supply voltage.

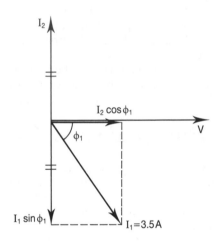

Fig 1.29

(b) From Fig. 1.29, it may be seen that
$I = I_1 \cos \phi_1$

so, $I = 3.5 \times 0.6$

hence, $I = 2.1$ A **Ans**

The following points should be noted. Firstly, by improving the overall load power factor as above, the current drawn from the supply has been considerably reduced (by 40%), yet the load power dissipation remains unchanged. However, it is not normal practice to correct the power factor to unity, because this results in the resonant condition. This is avoided in power circuits, since the current circulating between the capacitor and inductor can be many times the supply current. For this reason, in practical circuits, the power factor is not normally improved beyond 0.9. In addition, the closer we get to $\cos \phi = 1$, the larger the capacitor value required. Large value capacitors having a high working voltage are relatively expensive components. Consider the following example.

▪ Worked Example 1.11 ▪

 For the motor specified in the previous example (1.10), calculate (a) the value of capacitor required to improve the load power factor to 0.9, and (b) the current now drawn from the supply.

In order to solve this problem a phasor diagram is an important requirement. This phasor diagram is shown in Fig. 1.30.

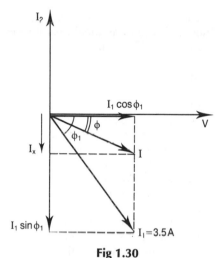

Fig 1.30

(a) From the phasor diagram:

$$I_1 \cos \phi_1 = 3.5 \times 0.6 = 2.1 \text{ A}$$

Also, considering the triangle consisting of I, $I_1 \cos \phi_1$ and I_x; where $\cos \phi = 0.9$:

$$\phi = \cos^{-1} 0.9 = 25.84°; \text{ so } \tan \phi = 0.4843$$

but $\tan \phi = \dfrac{I_x}{I_1 \cos \phi_1}$; so $I_x = I_1 \cos \phi_1 \times \tan \phi$

therefore, $I_x = 2.1 \times 0.4843 = 1.017 \text{ A}$

Now, $\qquad I_2 = I_1 \sin \phi_1 - I_x$

so $\qquad I_2 = (3.5 \times 0.8) - 1.017 = 1.783 \text{ A}$

$$X_C = \frac{V}{I_2} \text{ ohm} = \frac{240}{1.783} = 134.61 \ \Omega$$

and $\qquad C = \dfrac{1}{2\pi f X_C} \text{ farad} = \dfrac{1}{2 \times \pi \times 50 \times 134.61}$

hence $\qquad C = 23.65 \ \mu F$ **Ans**

(b) Once more, from the phasor diagram it may be seen that:

$$\cos \phi = \frac{I_1 \cos \phi_1}{I} \text{ so } I = \frac{I_1 \cos \phi_1}{\cos \phi}$$

hence $I = \dfrac{2.1}{0.9} = 2.33 \text{ A}$ **Ans**

From this example, it is clear that a smaller (and probably cheaper) capacitor is required, and the current drawn from the supply is still reduced considerably (by 33.3%). This example also illustrates the importance and usefulness of a phasor diagram.

1.12 Filters

A filter is a network designed to pass signals at certain frequencies, and to reject signals at all other frequencies. Ideally, a filter would introduce zero attenuation to selected frequencies, known as the pass band(s), and completely block all others (the stop band). Since a filter has to be frequency sensitive, then it must employ frequency dependent components; i.e. inductors or capacitors, or (usually) a combination of both of these components. There are four basic types of filter, namely, low-pass, high-pass, band-pass and band-stop. In addition, a filter circuit may be classified as being either passive or active. A passive filter is one which contains only a combination of resistors, inductors or capacitors, but no power source within it. An active filter consists of a combination of these components interconnected with an operational amplifier (0p. amp.). The latter would include its own power supply source, independent of the signal to be filtered.

1.13 Low-pass Filters

A very simple form of low-pass filter is shown in Fig. 1.31. The input voltage is shown as V_1 and

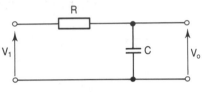

Fig 1.31

the output voltage as V_0. Since the reactance of the capacitor, $X_C \propto 1/f$, then at low frequencies, this reactance will be very high. Provided that the value of X_C is very much greater than the value of R, then V_0 will be almost the same value as V_1. The reason is that the circuit forms a potential divider circuit, such that $V_0 = [X_C/(R + X_C)] \times V_1$. As the frequency of the input is increased, so the value of X_C will decrease, but R remains constant. Thus, as f increases, so V_0 decreases. The frequency response curve for this simple arrangement is illustrated in Fig. 1.32. In this diagram, the

vertical axis is the *ratio* of the output voltage to the input voltage (V_0/V_1). The ideal response is shown by the dotted lines. The 'cut-off' frequency is shown as f_c.

So far, we have considered the behaviour of the circuit in terms of the input and output voltages. A similar argument applies if we consider the ratio of the input and output currents. Since at low frequencies X_C is high, virtually no current flows through the capacitor. Thus, the current reaching the output terminals is virtually the same as that entering at the input. At high frequencies, X_C is very low. In this case, the capacitor will divert (shunt) most of the current, thus preventing it from reaching the output terminals.

One problem with this form of circuit is the inclusion of the resistor. This will result in some unwanted attenuation of the signal in the low frequency pass band. One method of overcoming this problem is to use an active form of filter. In this case, the amplifier employed can make up for any *unwanted* attenuation. You are not required to explain the full circuit action of active filters, but a simplified circuit is shown in Fig. 1.33. The

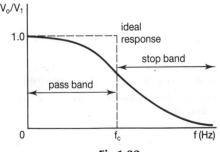

Fig 1.32

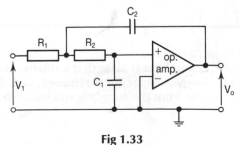

Fig 1.33

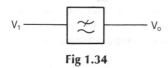

Fig 1.34

general block diagram symbol for a low-pass filter is shown in Fig. 1.34.

Another form of passive low-pass filter utilises capacitors and inductors. In order to minimise any unwanted attenuation, the inductors need to have as small a resistance as possible. These filters may be in the form of either a 'T' or a 'π' network. In each case, the series connected elements are the inductors, and the parallel elements are the capacitors. An example of a 'T' network low-pass

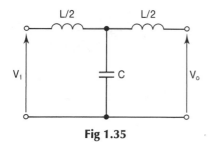

Fig 1.35

filter appears in Fig. 1.35. The 'π' type is shown in Fig. 1.36.

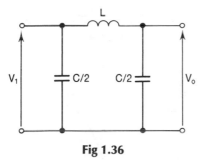

Fig 1.36

The action of both types is the same. At low frequencies, the reactance of the inductors (X_L) is low, since $X_L \propto f$. On the other hand, at low frequencies, the reactance of the capacitors will be very high. Hence, at these frequencies, the inductors offer little opposition to the signal, and the capacitors have negligible shunting effect. At high frequencies, the inductive reactance provides considerable attenuation of the signal. At the same time, the capacitors have considerable shunting effect. The frequency response curve for this type of filter is closer to the ideal than that of the simple R-C filter, and is illustrated in Fig. 1.37. One

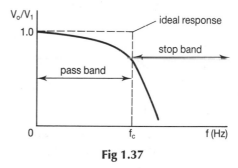

Fig 1.37

common example of the use of an L-C filter is the smoothing circuit employed between a full-wave bridge rectifier and the d.c. load being supplied. In this application, the filter is required to reduce the ripple in the waveform, thus resulting in a smooth d.c.

Note: The total inductance of inductors in series is given by their sum. Similarly, the total capacitance of two capacitors in parallel is given by their sum. Thus, in the circuits of Figs. 1.35 and 1.36, the total inductance and capacitance is L henry and C farads respectively.

1.14 High-pass Filters

The action of this filter is exactly the opposite to that of a low-pass filter. Ideally, it will stop all signals at frequencies below the cut-off value, and pass all frequencies above this value. A simple high-pass filter may also be achieved by the use of a C-R circuit, as shown in Fig. 1.38. At low frequencies, the high reactance of the capacitor

severely attenuates the signal, and only a very small proportion of the input voltage is developed (as the output) across the resistor. At high frequencies, the capacitive reactance is very low, and the output developed across the resistor is relatively large. The frequency response curve for this simple form of filter is illustrated in Fig. 1.39.

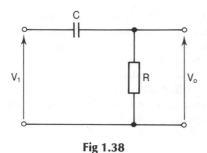

Fig 1.38

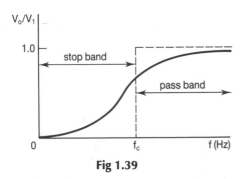

Fig 1.39

Figs. 1.40 and 1.41 show an active version of this filter, and the block diagram symbol for a high-pass filter, respectively.

This simple form of *C-R* filter suffers from the same disadvantages as its low-pass version. Thus it is more common to use the *L-C* types, in either 'T' or 'π' configuration. These are shown in Figs. 1.42 and 1.43, with the frequency response curve illustrated in Fig. 1.44. It is left for the reader to confirm the action of the circuit, bearing in mind

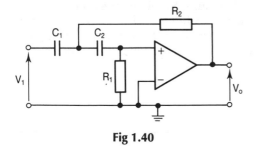

Fig 1.40

that the positions of capacitors and inductors in the circuit are reversed when compared with the corresponding low-pass versions.

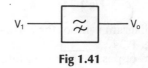

Fig 1.41

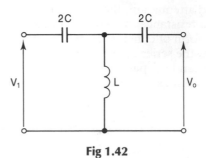

Fig 1.42

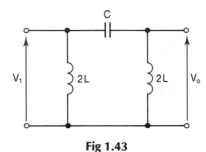

Fig 1.43

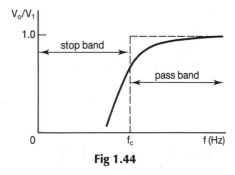

Fig 1.44

1.15 Band-pass Filters

This circuit is a combination of the *L-C* forms of the previous two types of filter circuits. It utilises both series and parallel tuned (resonant) circuits. The capacitor and inductor values are chosen so

that the same resonant frequency applies to both the series and parallel arrangements. The series impedance of the circuit is provided by two identical series tuned circuits. The shunt

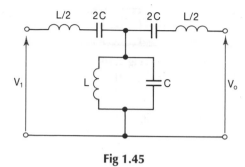

Fig 1.45

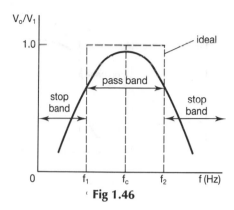

Fig 1.46

impedance is provided by the parallel tuned circuit, as shown in Fig. 1.45.

For signals at or near to f_0, the series impedance will be at its minimum possible value. The impedance of the parallel branch will be at its maximum value under these conditions. Thus, over a narrow band of frequencies, the opposition to the signal flow between input and output is minimal. At the same time, the shunting effect on the signal is also minimal. The attenuating effect of the whole circuit is therefore at its minimum.

At frequencies on either side of resonance, the series impedance is increased, whilst the shunt impedance is decreased. Hence, the signal flow between input and output is impeded, and the shunting effect is increased. The signal is thus attenuated as illustrated in Fig. 1.46. The selectivity, and hence bandwidth, of the circuit is determined by the ratio L/C. Fig. 1.47 shows the block diagram symbol for a band-pass filter.

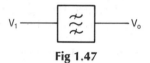

Fig 1.47

1.16 Band-stop (Notch) Filters

This filter circuit also utilises both series and parallel tuned elements. Compared with the band-pass filter, the roles of the series and parallel tuned elements are reversed. Thus, the series impedance is provided by two identical parallel-tuned elements. The shunt impedance is provided by a series-tuned element. Once more, the components are chosen so that a common resonant frequency is achieved. The circuit arrangement is shown in Fig. 1.48, and its action is as follows.

At or near to f_0, the series impedance will be at its maximum, and the shunt impedance will be at its minimum. Under these conditions maximum attenuation is achieved. At frequencies removed from resonance the attenuating effect of the circuit is reduced. This results in a frequency response curve as shown in Fig. 1.49. The band-stop filter symbol is as shown in Fig. 1.50.

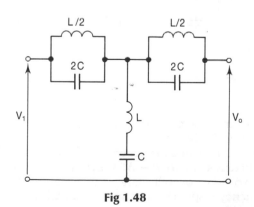

Fig 1.48

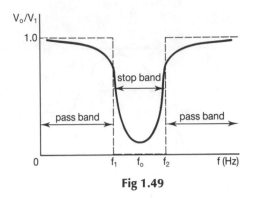

Fig 1.49

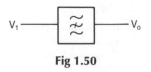

Fig 1.50

1.17 Coupled Tuned Circuits

In communication systems it would be ideal to
have a circuit having a frequency response curve
that has a flat top and vertical sides. Such a
response curve is shown in Fig. 1.46, as the ideal
response for a band-pass filter. This would allow
an amplifier to be designed having a large (and
uniform) gain over the required bandwidth, and
zero gain over all other frequencies. A close
approximation to this ideal curve may be achieved
by the use of coupled tuned circuits.

A coupled tuned circuit is obtained by having
two circuits of the same resonant frequency, and
weakly coupling them by a common element; e.g.
mutual inductance, self inductance or capacitance.
Let us consider one example of this, where the
coupling is provided by mutual inductance. A
typical circuit is shown in Fig. 1.51. This

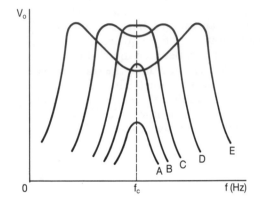

Fig 1.52

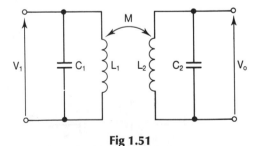

Fig 1.51

arrangement is also known as transformer
coupling. In this case, the degree of coupling
between the two coils depends upon their coupling
coefficient k, where $k = M/\sqrt{L_1 L_2}$ (see Volume 1,
Chap 5). The overall shape of the resulting
frequency response curve varies according to the

degree of coupling achieved, and is illustrated in
Fig. 1.52.

Curves A and B are the result of very loose
coupling. This may be defined as the case where
the coupling coefficient is less than the reciprocal
of the circuit Q factor, i.e. $k < 1/Q$. Since the
coupling is very loose, the impedance and response
curve for each individual tuned circuit is virtually
unaffected by the presence of the other. The
overall response curve is given by the product of
the two individual responses. Thus, the output
voltage tends to be small, but the selectivity is
high.

Curve C shows the overall response when
critical coupling is achieved ($k = 1/Q$). This
condition results in a flat top and steep sides for
the response curve. In this case, the output at the
centre frequency is larger than in any other. The

critically coupled circuit therefore approaches the ideal shape required.

If the coupling is increased beyond the critical value, so that $k > 1/Q$, then we have overcoupling. Under this condition the overall response curve splits into two separate peaks, as shown in curves D and E. The greater the coupling the wider the spacing of the peaks and the lower the dip between them.

Assignment Questions

1 A 100 Ω resistor is connected in parallel with a 10.61 μF capacitor, across a 120 V, 200 Hz supply. Determine (a) the current through each component, (b) the current drawn from the supply, and (c) the circuit impedance.

2 A 20 Ω resistor is connected in parallel with an inductor of negligible resistance, and inductance of 2.5 mH, across a 50 V, 1 kHz supply. Determine (a) the current in each branch, (b) the supply current, (c) the circuit impedance and, (d) the power consumed.

3 A coil of negligible resistance and inductance L henry is connected in parallel with a 50 Ω resistor, across a 250 V, 50 Hz supply. If the supply current is 10 A, determine (a) the current drawn by each component, (b) the value of the inductance, and (c) the circuit phase angle and power factor.

4 A 1.5 μF capacitor is connected in parallel with a 15 Ω resistor across a 12 V, 10 kHz supply. Calculate (a) the current in each branch, (b) the supply current, (c) the circuit impedance, (d) the power consumed, and (e) the values of apparent and reactive components of power.

5 A capacitor of C farads is connected in parallel with resistor of R ohms across a 40 V, 150 Hz supply. If the supply current is 0.5 A at a power factor of 0.8 leading, calculate the values of C and R.

6 A 75 mH inductor is connected in parallel with a 10 μF capacitor across a 50 V, 200 Hz supply. Determine (a) the branch currents, (b) the supply current, (c) the circuit impedance, and (d) the power dissipated.

7 A coil of resistance 50 Ω and inductance 320 mH is connected in parallel with a 15 μF capacitor, across a 250 V, 50 Hz supply. Calculate (a) the coil current, (b) the capacitor current, (c) the supply current, (d) the circuit phase angle and impedance, (e) the power consumed, and (f) the apparent and reactive components of power.

8 A coil of resistance 60 Ω and inductance 0.5 H is connected in parallel with a capacitor, across a 200 V, 50 Hz supply. If the capacitor draws a current of 942.5 mA, calculate (a) the capacitor value (b) the current through the coil, (c) the supply current (d) the circuit phase angle, and (e) the power consumed.

9 A coil of inductance 0.15 H and resistance 30 Ω is connected in series with a 47 nF capacitor, across a 200 mV, variable frequency supply. Determine (a) the frequency at which resonance occurs, (b) the circuit Q-factor, (c) the capacitor p.d. at resonance, and (d) the circuit current.

10 A 50 mH coil of resistance 15 Ω is connected in series with a variable capacitor, across a 6 V, 500 Hz supply. Calculate (a) the capacitor value that results in unity power factor for the circuit, (b) the resulting current, (c) the p.d. across the capacitor, and (d) the p.d. across the coil.

11 A 500 nF capacitor is connected in series with a coil, across a 5 V, 1.5 kHz supply. Under these conditions, the current has a value of 50 mA, and is in phase with the applied voltage. Determine (a) the resistance and inductance of the coil, (b) the circuit Q-factor, and (c) the circuit bandwidth.

12 A 5 μF capacitor is connected in parallel with a 10 mH inductor of resistance 2.5 Ω. Calculate (a) the resonant frequency, (b) the resonant frequency if an additional 22 Ω resistor is connected in series with the coil, and (c) the Q-factor in each case.

13 A coil of inductance 100 mH and resistance 400 Ω is connected in parallel with a 10 nF capacitor, across a 12 V a.c. supply. Determine, for the condition when the supply current is a minimum, (a) the frequency, (b) the circuit impedance, (c) the supply current, and (d) the Q-factor.

14 A capacitor is connected in parallel with a coil of inductance 300 mH and resistance 10 Ω. Resonance occurs when the supply is 40 V at a frequency of 2 kHz. Calculate, (a) the capacitor value, (b) the supply current, (c) the circuit Q-factor, and (d) the circuit bandwidth.

15 A 5 kW motor has a power factor of 0.6 lagging. When it is connected to a 240 V, 50 Hz supply, determine (a) the current drawn from the supply, (b) the value of power factor correction capacitor to produce unity power factor, (c) the supply current under this condition.

Explain why unity power factor would normally be avoided for this type of circuit.

16 For the motor and supply specified in Question 15 above, calculate (a) the value of capacitor required to improve the original power factor to 0.9 lagging, (b) this capacitor's VAr rating, and (c) the current now drawn from the supply.

17 A 240 V, 50 Hz supply feeds the following loads: (i) incandescent lamps taking a current of 6.5 A at unity power factor; (ii) fluorescent lamps taking a current of 8 A at a power factor of 0.9 leading; and (iii) a motor taking a current of 15 A at a lagging power factor of 0.625. Determine (a) the total current drawn from the supply, and the overall power factor, and (b) the value of power factor correction capacitor required to improve the overall power factor to unity.

Suggested Practical Assignments

Note: Component values and specific items of equipment when quoted here are only suggestions. Those used in practice will naturally depend upon availability within a given institution.

Assignment 1

To investigate the effect of parallel resonance.

Apparatus:

1 × signal generator
1 × inductor (of known value)
1 × capacitor (of known value)
1 × voltmeter
1 × ammeter

Method:

1 Connect the circuit of Fig. 1.53, and calculate the theoretical resonant frequency (f_0) of the circuit.

2 Set the signal generator frequency to $f_0/10$, and record this, together with the values for the applied voltage and circuit current.

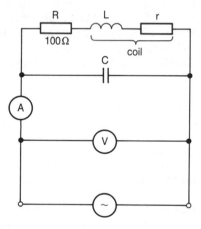

Fig 1.53

3 Ensuring that the applied voltage is maintained constant throughout, increase the frequency, in discrete steps, up to $f_0 \times 10$. Record the current at each step.
   Note: Take 'extra' readings at, and on either side of the resonant frequency.

4 From your recorded values, calculate and tabulate the corresponding values of circuit impedance.

5 Plot (on the same axes) graphs of circuit current and impedance, versus frequency.

6 From your graphs, determine the actual resonant frequency of the circuit, and also determine the resistance value of the inductor.

7 Complete an assignment report. This should include the circuit diagram, a description of the procedure carried out, all tabulated readings, calculations, and conclusions drawn.

Assignment 2

To observe the effects of power factor correction.

Apparatus:
1 × 0.1 H coil
1 × variable capacitor, 0.1 to 10 μF
1 × 100 Ω resistor
1 × double beam oscilloscope
1 × ammeter
1 × a.c. signal generator

Method:

1 Connect the circuit as in Fig. 1.54.

2 Set the signal generator to 20 V, at a frequency of 500 Hz.

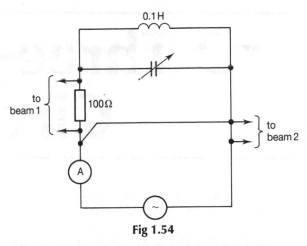

Fig 1.54

3 Disconnect the capacitor, and measure the circuit current. Note the phase angle of the circuit by comparing the two waveforms on the oscilloscope.

4 Set the capacitor to 0.1 μF, and reconnect it to the circuit.

5 Measure the circuit current and monitor the phase angle on the oscilloscope.

6 Repeat the procedure of paragraph (5) above, for increasing values of capacitance.

7 Complete an assignment report.

▪ 2 Three-Phase a.c. Circuits ▪

This chapter introduces the concepts and principles of the three-phase electrical supply, and the corresponding circuits. On completion you should be able to:

1] Describe the reasons for, and the generation of the three-phase supply.

2] Distinguish between star (3 and 4-wire) and delta connections.

3] State the relative advantages of three-phase systems compared with single-phase systems.

4] Solve three-phase circuits in terms of phase and line quantities, and the power developed in three-phase balanced loads.

5] Measure power dissipation in both balanced and unbalanced three-phase loads, using the 1, 2 and 3-wattmeter methods, and hence determine load power factor.

6] Calculate the neutral current in a simple unbalanced 4-wire system.

2.1 Generation of a Three-Phase Supply

In order to understand the reasons for, and the method of generating a three-phase supply, let us firstly consider the generation of a single-phase supply. Alternating voltage is provided by an a.c. generator, more commonly called an *alternator*. The basic principle was outlined in Volume 1, Chapter 5. It was shown that when a coil of wire, wound on to a rectangular former, is rotated in a magnetic field, an alternating (sinusoidal) voltage is induced into the coil. You should also be aware that for electromagnetic induction to take place, it is the *relative* movement between conductor and magnetic flux that matters. Thus, it matters not whether the field is static and the conductor moves, or vice versa.

In this context, the term **field** refers to the magnetic field. This field is normally produced by passing d.c. current through the rotor winding. Since the winding is rotated, the current is passed to it via copper slip-rings on the shaft. The external d.c. supply is connected to the slip-rings by a pair of carbon brushes.

For a practical alternator it is found to be more convenient to rotate the magnetic field, and to keep the conductors (coil or winding) stationary.

In any rotating a.c. machine, the rotating part is called the rotor, and the stationary part is called the stator. Thus, in an alternator, the **field** system is contained in the rotor. The winding in which the emf is generated is contained in the stator. The reasons for this are as follows:

(a) When large voltages are generated, heavy insulation is necessary. If this extra mass has to be rotated, the driving device has to develop extra power. This will then reduce the overall efficiency of the machine. Incorporating the winding in the stator allows the insulation to be as heavy as necessary, without adversely affecting the efficiency.

(b) The contact resistance between the brushes and slip-rings is very small. However, if the alternator provided high current output (in hundreds of ampere), the I^2R power loss would be significant. The d.c. current (excitation current) for the field system is normally only a few amps or tens of amps. Thus, supplying the field current via the slip-rings produces minimal power loss. The stator winding is simply connected to terminals on the outside of the stator casing.

(c) For very small alternators, the rotor would contain permanent magnets to provide the rotating field system. This then altogether eliminates the need for any slip-rings. This arrangement is referred to as a brushless machine.

The basic construction for a single-phase alternator is illustrated in Fig. 2.1. The conductors

of the stator winding are placed in slots around the inner periphery of the stator. The two ends of this winding are then led out to a terminal block on the casing. The rotor winding is also mounted in slots, around the circumference of the rotor. This figure is used to illustrate the principle. A practical machine would have many more conductors and slots.

Since the conductors of the stator winding are spread around the whole of the slots, it is known as a distributed winding. As the rotor field sweeps past these conductors an emf is induced in each of them in turn. These individual emfs reach their maximum values only at the instant that the rotating field 'cuts' them at 90°. Also, since the slots have an angular displacement between them, then the conductor emfs will be out of phase with each other by this same angle. In Fig. 2.1 there are a total of twelve conductors, so this phase difference must be 30°. The total stator winding emf will therefore not be the *arithmetic* sum of the conductor emfs, but will be the *phasor* sum, as shown in Fig. 2.2. The ratio of the phasor sum to

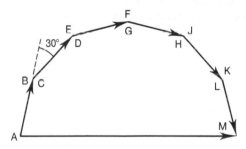

AB, CD etc are conductor (coil) emfs.
AM is the phasor sum

Fig 2.2

the arithmetic sum is called the distribution factor. For the case shown (a fully distributed winding) the distribution factor is 0.644.

Now, if all of the stator conductors could be placed into a single pair of slots, opposite to each other, then the induced emfs would all be in phase. Hence the phasor and arithmetic sums would be the same, yielding a distribution factor of unity. This is not a practical solution. However, if the conductors are concentrated so as to occupy only one third of the available stator slots, then the distribution factor becomes 0.966. In a practical single-phase alternator, the stator winding is distributed over two thirds of the slots.

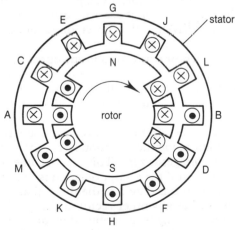

Fig 2.1

Let us return to the option of using only one third of the slots. We will now have the space to put two more identical windings into the stator. Each of the three windings could be kept electrically separate, with their own pairs of terminals. We would then have three separate single-phase alternators in the same space as the original. Each of these would also have a good distribution factor of 0.966. The three winding emfs will of course be mutually out of phase with each other by 120°, since each whole winding will occupy 120° of stator space. What we now have is the basis of a three-phase alternator.

The term three-phase alternator is in some ways slightly misleading. What we have, in effect, are three identical single-phase alternators contained in the one machine. The three stator windings are brought out to their own separate pairs of terminals on the stator casing. These stator windings are referred to as phase windings, or phases. They are identified by the colours red, yellow and blue. Thus we have the red, yellow and blue phases. The circuit representation for the stator winding of such a machine is shown in Fig. 2.3. In this figure, the three phase windings are

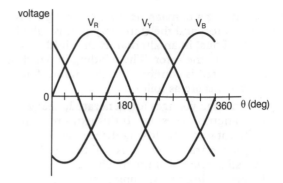

voltage

Fig 2.4

and phasor diagrams are shown in Figs. 2.4 and 2.5 respectively. In either case, we need to select a reference phasor. By convention, the reference is always taken to be the red phase voltage. The

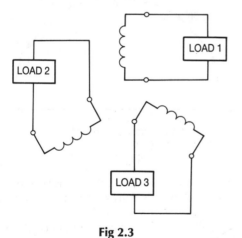

Fig 2.3

shown connected, each one to its own separate load. This arrangement is known as a three-phase, six-wire system. However, three-phase alternators are rarely connected in this way.

Since the three generated voltages are sinewaves of the same frequency, mutually out-of-phase by 120°, then they may be represented both on a waveform diagram using the same angular or time axis, and as phasors. The corresponding waveform

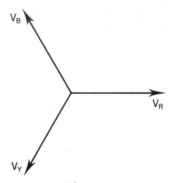

Fig 2.5

yellow phase lags the red by 120°, and the blue lags the red by 240° (or, if you prefer, leads the red by 120°). The windings are arranged so that when the rotor is driven in the chosen direction, the phase sequence is red, yellow, blue. If, for any reason, the rotor was driven in the opposite direction, then the phase sequence would be reversed, i.e. red, blue, yellow. We shall assume that the normal sequence of R, Y, B applies at all times.

It may be seen from the waveform diagram that, at any point along the horizontal axis, the sum of the three voltages is zero. This fact becomes even more apparent if the phasor diagram is redrawn as in Fig. 2.6. In this diagram, the three phasors have been treated as any other vector quantity. The sum of the vectors may be determined by drawing them to scale, as in Fig. 2.6, and the resultant found by

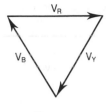

Fig 2.6

measuring the distance and angle from the beginning point of the first vector to the arrowhead of the last one. If, as in Fig. 2.6, the first and last vectors meet in a closed figure, the resultant must be zero.

2.2 Three-phase, Four-wire System

It is not necessary to have six wires from the three phase windings to the three loads, provided there is a common 'return' line. Each winding will have a 'start' (S) and a 'finish' (F) end. The common connection mentioned above is achieved by connecting the corresponding ends of the three phases together. For example, either the three 'F' ends or the three 'S' ends are commoned. This form of connection is shown in Fig. 2.7, and is

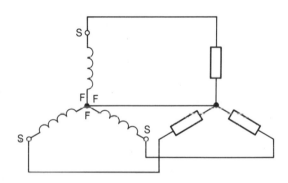

Fig 2.7

known as a *star* or Y connection. With the resulting 4-wire system, the three loads also are connected in star configuration. The three outer wires are called the *lines*, and the common wire in the centre is called the *neutral*.

If the three loads were identical in every way (same impedance and phase angle), then the currents flowing in the three lines would be identical. If the waveform and/or phasor diagrams

for these currents were drawn, they would be identical in form to Figs. 2.4 and 2.5. These three currents meet at the star point of the load. The resultant current returning down the neutral wire would therefore be zero. The load in this case is known as a balanced load, and the neutral is not strictly necessary. However it is difficult, in practice, to ensure that each of the three loads are exactly balanced. For this reason the neutral is left in place. Also, since it has to carry only the relatively small 'out-of-balance' current, it is made half the cross-sectional area of the lines.

Let us now consider one of the advantages of this system compared with both a single-phase system, and the three-phase 6-wire system. Suppose that three identical loads are to be supplied with 200 A each. The two lines from a single-phase alternator would have to carry the total 600 A required. If a 3-phase, 6-wire system was used, then each line would have to carry only 200 A. Thus, the conductor csa would only need to be 1/3 that for the single-phase system, but of course, being six lines would entail using the same total amount of conductor material. If a 4-wire, 3-phase system is used there will be a saving on conductor costs in the ratio of 3.5 : 6 (the 0.5 being due to the neutral). If the power has to be sent over long transmission lines, such as the National Grid System, then the 3-phase, 4-wire system yields an enormous saving in cable costs. This is one of the reasons why the power generating companies use three-phase, star-connected generators to supply the grid system.

2.3 Relationship between Line and Phase Quantities in a Star-connected System

Consider Fig. 2.8, which represents the stator of a 3-phase alternator connected to a 3-phase balanced load. The voltage generated by each of the three phases is developed between the

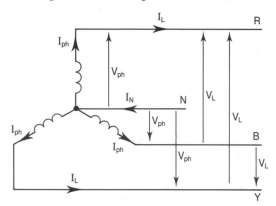

Fig 2.8

appropriate line and the neutral. These are called the phase voltages, and may be referred to in general terms as V_{ph}, or specifically as V_{RN}, V_{YN} and V_{BN} respectively. However, there will also be a difference of potential between any pair of lines. This is called a line voltage, which may be generally referred to as V_L, or specifically as V_{RY}, V_{YB} and V_{BR} respectively.

A line voltage is the *phasor difference* between the appropriate pair of phase voltages. Thus, V_{RY} is the phasor difference between V_{RN} and V_{YN}. In terms of a phasor diagram, the simplest way to subtract one phasor from another is to reverse one of them, and then find the resulting phasor sum. This is, mathematically, the same process as saying that $a - b = a + (-b)$. The corresponding phasor diagram is shown in Fig. 2.9.

Note: If V_{YN} is reversed, it is denoted either as $-V_{YN}$ or as V_{NY}. We shall use the first of these.

The phasor difference between V_{RN} and V_{YN} is simply the phasor sum of $V_{RN} + (-V_{YN})$. Geometrically this is obtained by completing the parallelogram as shown in Fig. 2.9. This parallelogram consists of two isosceles triangles, such as OCA. Another property of a parallelogram is that its diagonals bisect each other at right angles. Thus, triangle OCA consists of two equal

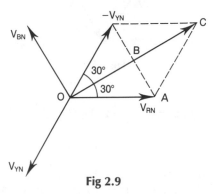

Fig 2.9

right-angled triangles, OAB and ABC. This is illustrated in Fig. 2.10. Since triangle OAB is a 30°, 60°, 90° triangle, then the ratios of its sides AB:OA:OB will be $1:2:\sqrt{3}$ respectively.

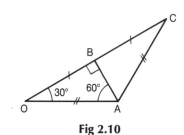

Fig 2.10

Hence, $\dfrac{OB}{OA} = \dfrac{\sqrt{3}}{2}$

so $OB = \dfrac{\sqrt{3}.OA}{2}$

but $OC = 2 \times OB = \sqrt{3}.OA$

and since $OC = V_{RY}$, and $OA = V_{RN}$

then $V_{RY} = \sqrt{3}\ V_{RN}$

Using the same technique, it can be shown that:

$V_{YB} = \sqrt{3}\ V_{YN}$ and $V_{BR} = \sqrt{3}\ V_{BN}$

Thus, in star configuration, $V_L = \sqrt{3}V_{ph}$ (2.1)

The complete phasor diagram for the line and phase voltages for a star connection is shown in Fig. 2.11. Also, considering the circuit diagram of

Fig. 2.8, the line and phase currents must be the same.

Hence, in star configuration, $I_L = I_{ph}$ (2.2)

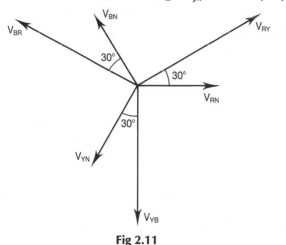

Fig 2.11

We now have another advantage of a 3-phase system compared with single-phase. The star-connected system provides two alternative voltage outputs from a single machine. For this reason, the stators of all alternators used in electricity power stations are connected in star configuration. These machines normally generate a line voltage of about 25 kV. By means of transformers, this voltage is stepped up to 400 kV for long distance transmission over the National Grid. For more localised distribution, transformers are used to step down the line voltage to 132 kV, 33 kV, 11 kV, and 415 V. The last three of these voltages are supplied to various industrial users. The phase voltage derived from the 415 V lines is 240 V, and is used to supply both commercial premises and households.

▪ Worked Example 2.1 ▪

 A 415 V, 50 Hz, 3-phase supply is connected to a star-connected balanced load. Each phase of the load consists of a resistance of 25 Ω and inductance 0.1 H, connected in series. Calculate (a) phase voltage, (b) the line current drawn from the supply, and (c) the power dissipated.

A

Whenever a three-phase supply is specified, the voltage quoted is always the line voltage. Also, since we are dealing with a balanced load, then it necessary only to calculate values for one phase of the load. The figures for the other two phases and lines will be identical to these.

$V_L = 415$ V; $f = 50$ Hz; $R_{ph} = 25$ Ω; $L_{ph} = 0.1$ H

(a)
$$V_{ph} = \frac{V_L}{\sqrt{3}} = \frac{415}{\sqrt{3}}$$

so $V_{ph} = 240$ V **Ans**

(b) Since it is possible to determine the impedance of a phase of the load, and we

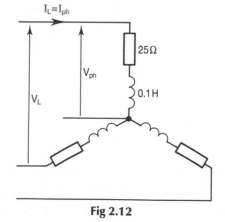

Fig 2.12

now know the phase voltage, then the phase current may be calculated:

$$X_L = 2\pi fL \text{ ohm} = 2\pi \times 50 \times 0.1$$

hence $X_L = 31.42 \ \Omega$

$$Z_{ph} = \sqrt{R_{ph}^2 + X_L^2} \text{ ohm} = \sqrt{25^2 + 31.42^2}$$

$$Z_{ph} = 40.15 \ \Omega$$

$$I_{ph} = \frac{V_{ph}}{Z_{ph}} \text{ amp} = \frac{240}{40.15}$$

so $I_{ph} = 5.98$ A

In a star-connected circuit, $I_L = I_{ph}$

therefore $I_L = 5.98$ A **Ans**

The power in one phase, $P_{ph} = I_{ph}^2 R_{ph}$ watt
$$= 5.98^2 \times 25$$
$$P_{ph} = 893.29 \text{ W}$$

and since there are three phases, then the total power is:

$$P = 3 \times P_{ph} \text{ watt} = 3 \times 893.29$$

hence $P = 2.68$ kW **Ans**

2.4 Delta or Mesh Connection

If the start end of one winding is connected to the finish end of the next, and so on until all three windings are interconnected, the result is the delta or mesh connection. This connection is shown in Fig. 2.13. The delta connection is not reserved for machine windings only, since a 3-phase load may also be connected in this way.

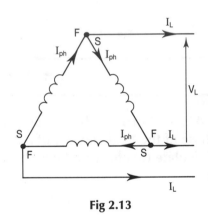

Fig 2.13

2.5 Relationship between Line and Phase Quantities in a Delta-connected System

It is apparent from Fig. 2.13 that each pair of lines is connected across a phase winding. Thus, for the delta-connection:

$$V_L = V_{ph} \qquad (2.3)$$

It is also apparent that the current along each line is the phasor difference of a pair of phase currents. The three phase currents are mutually displaced by 120°, and the phasor diagram for these is shown in Fig. 2.14. Using exactly the same geometrical technique as that for the phase and line voltages in the star connection, it can be shown that:

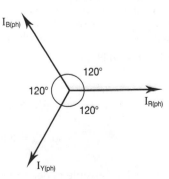

Fig 2.14

$$I_L = \sqrt{3}\, I_{ph} \qquad (2.4)$$

The phasor diagram for the phase and line currents in delta connection is as in Fig. 2.15. Note that the provision of a neutral wire is not applicable with a delta connection. However, provided that the load is balanced, there is no requirement for one. Under balanced load conditions the three phase currents will be equal, as will be the three line currents. If the load is unbalanced, then these equalities do not exist, and each phase or line current would have to be calculated separately. This technique is beyond the scope of the syllabus you are now studying.

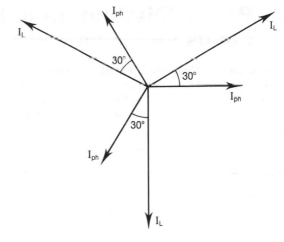

Fig 2.15

■ **Worked Example 2.2** ■

Q A balanced load of phase impedance 120 Ω is connected in delta. When this load is connected to a 600 V, 50 Hz, 3-phase supply, determine (a) the phase current, and (b) the line current drawn.

$Z_{ph} = 120\ \Omega;\ V_L = 600\ V;\ f = 50\ Hz$

The circuit diagram is shown in Fig. 2.16.

(a)   In delta, $V_{ph} = V_L = 600\ V$

$$I_{ph} = \frac{V_{ph}}{Z_{ph}}\ amp = \frac{600}{120}$$

so, $I_{ph} = 5\ A$ **Ans**

(b)   in delta, $I_L = \sqrt{3}\, I_{ph}$ amp $= \sqrt{3} \times 5$

therefore $I_L = 8.66\ A$ **Ans**

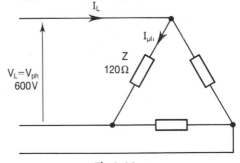

Fig 2.16

2.6 Power Dissipation in Star and Delta-connected Loads

We have seen in Example 2.1 that the power in a 3-phase balanced load is obtained by multiplying the power in one phase by 3. In many practical situations, it is more convenient to work with line quantities.

$$P_{ph} = V_{ph}I_{ph} \cos \phi \text{ watt}$$

where $\cos \phi$ is the phase power factor.

total power, $P = 3 \times P_{ph}$

$\qquad = 3 \times V_{ph}I_{ph} \cos \phi$ watt ... [1]

Considering a STAR-connected load,

$$V_{ph} = \frac{V_L}{\sqrt{3}}, \text{ and } I_{ph} = I_L$$

Substituting for V_{ph} and I_{ph} in equ [1]:

$$P = 3 \times \frac{V_L}{\sqrt{3}} I_L \cos \phi$$

therefore $P = \sqrt{3} \ V_L I_L \cos \phi$ watt $\qquad$ (2.4)

For a delta-connected load,

$$V_{ph} = V_L \text{ and } I_{ph} = \frac{I_{ph}}{\sqrt{3}}$$

and substituting these values into equ [1] will yield the same result as shown in (2.4) above. Thus, the *equation* for determining power dissipation, in both star and delta-connected loads is exactly the same. However, the *value* of power dissipated by a given load when connected in star is not the same as when it is connected in delta. This is demonstrated in the following example.

■ Worked Example 2.3 ■

  A balanced load of phase impedance 100 Ω and power factor 0.8 is connected (a) in star, and (b) in delta, to a 400 V, 3-phase supply. Calculate the power dissipation in each case.

$Z_{ph} = 100 \ \Omega$; $\cos \phi = 0.8$; $V_L = 400$ V

(a) the circuit diagram is shown in Fig. 2.17

$$V_{ph} = \frac{V_L}{\sqrt{3}} \text{ volt} = \frac{400}{\sqrt{3}}$$

so $V_{ph} = 231$ V

$$I_L = I_{ph} = \frac{V_{ph}}{Z_{ph}} \text{ amp} = \frac{231}{100}$$

and $I_L = 2.31$ A

$\qquad P = \sqrt{3} \ V_L I_L \cos \phi$ watt $= \sqrt{3} \times 400 \times 2.31 \times 0.8$

therefore $P = 1.28$ kW **Ans**

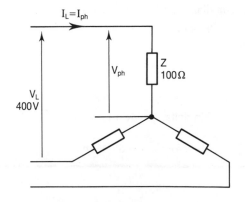

Fig 2.17

(b)   The circuit diagram is shown in Fig. 2.18

$V_{ph} = V_L = 400$ V

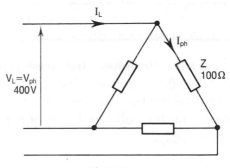

$$\frac{V_{ph}}{Z_{ph}} \text{ amp} = \frac{400}{100}$$

so $I_{ph} = 4$ A

but $I_L = \sqrt{3}\, I_{ph} = \sqrt{3} \times 4$

so $I_L = 6.93$ A

$P = \sqrt{3}\ V_L I_L \cos\phi \text{ watt} = \sqrt{3} \times 400 \times 6.93 \times 0.8$

therefore $P = 3.84$ kW **Ans**

Fig 2.18

Comparing the two answers for the power dissipation in the above example, it may be seen that:

Power in a delta-connected load is *three times* that when it is connected in star configuration.   (2.5)

2.7 Star/Delta Supplies and Loads

As explained in Section 2.3, the distribution of 3-phase supplies is normally at a much higher line voltage than that required for many users. Hence, 3-phase transformers are used to step the voltage down to the appropriate value. A three-phase transformer is basically three single-phase transformers interconnected. The three primary windings may be connected either in star or delta, as can the three secondary windings. Similarly, the load connected to the transformer secondary windings may be connected in either configuration. One important point to bear in mind is that the transformation ratio (voltage or turns ratio) refers to the ratio between the primary *phase* to the secondary *phase* winding. The method of solution of this type of problem is illustrated in the following worked example.

■ Worked Example 2.4 ■

Q Figure 2.19 shows a balanced, star-connected load of phase impedance 25 Ω and power factor 0.75, supplied from the delta-connected secondary of a 3-phase transformer. The turns ratio of the transformer is 20:1, and the star-connected primary is supplied at 11 kV. Determine (a) the voltages V_2, V_3 and V_4, (b) the currents I_1, I_2, and I_3 and (c) the power drawn from the supply.

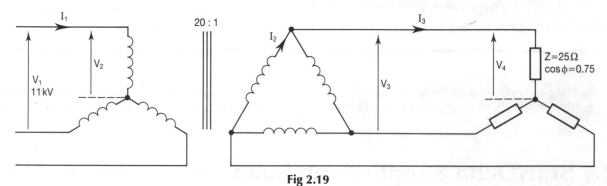

Fig 2.19

A ─────────────────────────────────────

$V_1 = 11000$ V; $N_p/N_s = 20/1$; $Z_{ph} = 25$ Ω; cos φ = 0.75

(a)

$$V_2 = \frac{V_1}{\sqrt{3}} = \frac{11000}{\sqrt{3}}$$

$$V_2 = 6.351 \text{ kV Ans}$$

$$\frac{V_3}{V_2} = \frac{N_s}{N_p}$$

so $$V_3 = \frac{N_s V_2}{N_p} \text{ volt} = \frac{6351}{20}$$

hence $V_3 = 317.45$ V **Ans**

$$V_4 = \frac{V_3}{\sqrt{3}} = \frac{317.45}{\sqrt{3}}$$

and $V_4 = 183.3$ V **Ans**

(b) In order to calculate the currents, we shall have to start with the load, and work 'back' through the circuit to the primary of the transformer:

$$I_3 = \frac{V_4}{Z_{ph}} \text{ amp} = \frac{183.3}{25}$$

$I_3 = 7.33$ A **Ans**

$$I_2 = \frac{I_3}{\sqrt{3}} = \frac{7.33}{\sqrt{3}}$$

$I_2 = 4.23$ A **Ans**

$$\frac{I_1}{I_2} = \frac{N_s}{N_p}$$

so $$I_1 = \frac{N_s I_2}{N_p} \text{ amp} = \frac{4.23}{20}$$

hence $I_1 = 0.212$ A **Ans**

(c) $P = \sqrt{3}\ V_1 I_1 \cos φ$ watt
$= \sqrt{3} \times 11 \times 10^3 \times 0.212 \times 0.75$

therefore $P = 3.02$ kW **Ans**

2.8 Measurement of Three-phase Power

In an a.c. circuit the true power may only be measured directly by means of a wattmeter. The principle of operation of this instrument is described in Volume 1, Chapter 5. As a brief reminder, the instrument has a fixed coil through which the load current flows, and a moving voltage coil (or pressure coil) connected in parallel with the load. The deflection of the pointer, carried by the moving coil, automatically takes into account the phase angle (or power factor) of the load. Thus the wattmeter reading indicates the true power, $P = VI \cos \phi$ watt.

If a three-phase load is balanced, then it is necessary only to measure the power taken by one phase. The total power of the load is then obtained by multiplying this figure by three. This technique can be very simply applied to a balanced, star-connected system, where the star point and/or the neutral line are easily accessible. This is illustrated in Fig. 2.20.

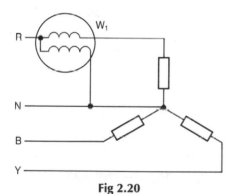

Fig 2.20

In the situation where the star point is not accessible, then an artificial star point needs to be created. This is illustrated in Fig. 2.21, where the value of the two additional resistors is equal to the resistance of the wattmeter voltage coil.

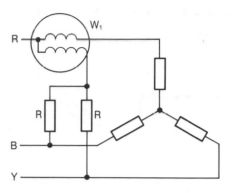

Fig 2.21

In the case of an unbalanced star-connected load, one or other of the above procedures would have to be repeated for each phase in turn. The total power $P = P_1 + P_2 + P_3$, where P_1, P_2, and P_3 represent the three separate readings.

For a delta-connected load, the procedure is not quite so simple. The reason is that the phase current is not the same as the line current. Thus, if possible, one of the phases must be disconnected to allow the connection of the wattmeter current coil. This is shown in Fig. 2.22. Again, if the load was unbalanced, this process would have to be repeated for each phase.

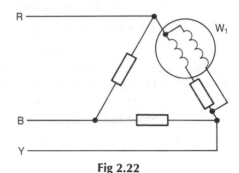

Fig 2.22

2.9 The Two-Wattmeter Method

The measurement of three-phase power using the above methods can be very awkward and time-consuming. In practical circuits, the power is usually measured by using two wattmeters simultaneously, as shown in Fig. 2.23.

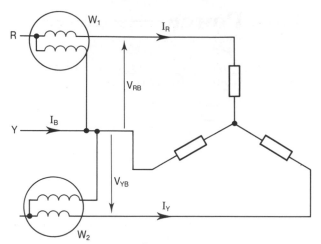

Fig 2.23

The advantages of this method are:

(a) Access to the star point is not required.

(b) The power dissipated in both balanced and unbalanced loads is obtained, without any modification to the connections.

(c) For balanced loads, the power factor may be determined.

Considering Fig. 2.23, the following statements apply:

Instantaneous power for W_1, $p_1 = v_{RB}i_R$ watt

and for W_2, $p_2 = v_{YB}i_Y$ watt

total instantaneous power = $p_1 + p_2$

$$= v_{RB}i_R + v_{YB}i_Y \quad [1]$$

Now, any line voltage is the phasor difference between the appropriate pair of phase voltages, hence:

$$v_{RB} = v_{RN} - v_{BN} \text{ and } v_{YB} = v_{YN} - v_{BN}$$

and substituting these into equ. [1] yields:

$$p_1 + p_2 = i_R(v_{RN} - v_{BN}) + i_Y(v_{YN} - v_{BN})$$

$$= v_{RN}i_R + v_{YN}i_Y - v_{BN}(i_R + i_Y)$$

but, $\quad i_R + i_Y = -i_B$

i.e. the phasor sum of three equal currents is zero.

therefore $p_1 + p_2 = v_{RN}i_R + v_{YN}i_Y + v_{BN}i_B$

The instantaneous sum of the powers measured by the two wattmeters is equal to the sum of instantaneous power in the three phases.

Hence, total power,

$$P = P_1 + P_2 = P_R + P_Y + P_B \text{ watt} \quad (2.6)$$

Consider now the phasor diagram for a resistive-inductive balanced load, with the two wattmeters connected as in Fig. 2.23. This phasor diagram appears as Fig. 2.24, below.

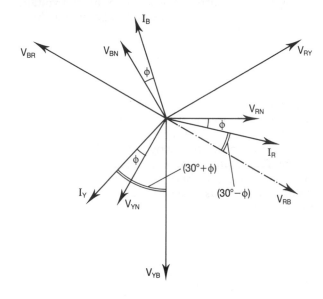

Fig 2.24

The power indicated by W_1,

$$p_1 = V_L I_L \cos(30° - \phi) \quad (2.7)$$

and that for W_2,

$$p_2 = V_L I_L \cos(30° + \phi) \quad (2.8)$$

From these results, and using Fig. 2.25, the following points should be noted:

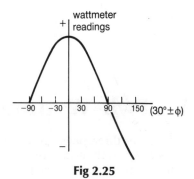

Fig 2.25

1 If the load p.f. > 0.5 (i.e. $\phi < 60°$); both meters will give a positive reading.

2 If the load p.f. = 0.5 (i.e. $\phi = 60°$); W_1 indicates the total power, and W_2 indicates zero.

3 If the load p.f. < 0.5 (i.e. $\phi > 60°$); W_2 attempts to indicate a negative reading. In this case, the connections to the voltage coil of W_2 need to be reversed, and the resulting reading recorded as a negative value. Under these circumstances, the total load power will be $P = P_1 - P_2$.

4 The load power factor may be determined from the two wattmeter readings from the equation:

$$\phi = \tan^{-1}\sqrt{3}\left(\frac{P_2 - P_1}{P_2 + P_1}\right) \qquad (2.9)$$

hence, power factor, cos ϕ can be determined.

▪ Worked Example 2.5 ▪

Q The power in a 3-phase balanced load was measured, using the two-wattmeter method. The recorded readings were 3.2 kW and 5 kW respectively. Determine the load power and power factor.

A ────────────────────────────

$P_1 = 5$ kW; $P_2 = 3.2$ kW

$$P = P_1 + P_2 \text{ watt} = (3.2 + 5) \text{ kW}$$

therefore, $P = 8.2$ kW **Ans**

$$\phi = \tan^{-1}\sqrt{3}\left(\frac{P_2 - P_1}{P_2 + P_1}\right)$$

$$= \tan^{-1}\sqrt{3}\left(\frac{5 - 3.2}{5 + 3.2}\right)$$

hence, $\phi = 20.82°$

and p.f. = cos ϕ = 0.935 **Ans**

■ Worked Example 2.6 ■

Q A 3-phase balanced load takes a line current of 24 A at a lagging power factor of 0.42, when connected to a 415 V, 50 Hz supply. If the power dissipation is measured using the two-wattmeter method, determine the two wattmeter readings, and the value of power dissipated. Comment on the results.

A

$I_L = 24$ A; $\cos \phi = 0.42$; $V_L = 415$ V

$$\phi = \cos^{-1} 0.42 = 65.17°$$

$$P_1 = V_L I_L \cos(30° - \phi) \text{ watt}$$

$$= 415 \times 24 \times \cos(-35.17°)$$

therefore, $P_1 = 8.142$ kW **Ans**

$$P_2 = V_L I_L \cos(30° + \phi) \text{ watt}$$

$$= 415 \times 24 \times \cos(95.17°)$$

therefore, $P_2 = -896.7$ W **Ans**

$$P = P_1 + P_2 \text{ watt} = 8142 + (-896.7)$$

hence, $P = 7.244$ kW **Ans**

To obtain the negative reading on W_2, the connections to its voltage coil must have been reversed.

2.10 Neutral Current in an Unbalanced Three-phase Load

We have seen that the neutral current for a balanced load is zero. This is because the phasor sum of three equal currents, mutually displaced by 120°, is zero. If the load is unbalanced, then the three line (and phase) currents will be unequal. In this case, the neutral has to carry the resulting out-of-balance current. This current is simply obtained by calculating the phasor sum of the line currents.

The technique is basically the same as that used previously, by resolving the phasors into horizontal and vertical components, and applying Pythagoras' theorem. The only additional fact to bear in mind is that both horizontal and vertical components can have negative values. This is illustrated by the following example.

▪ **Worked Example 2.7** ▪

Q An unbalanced, star-connected load is supplied from a 3-phase, 415 V source. The three phase loads are purely resistive. These loads are 25 Ω, 30 Ω and 40 Ω, and are connected in the red, yellow and blue phases respectively. Determine the value of the neutral current, and its phase angle relative to the red phase current.

A _____

$V_L = 415$ V; $R_R = 25$ Ω; $R_Y = 30$ Ω; $R_B = 40$ Ω

The circuit diagram is shown in Fig. 2.26.

$$V_{ph} = \frac{V_L}{\sqrt{3}} \text{ volt} = \frac{415}{\sqrt{3}} = 240 \text{ V}$$

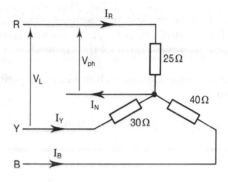

Fig 2.26

$$I_R = \frac{V_{ph}}{R_R} \text{ amp}; \ I_Y = \frac{V_{ph}}{R_Y} \text{ amp}; \ I_B = \frac{V_{ph}}{R_B} \text{ amp}$$

$$= \frac{240}{25} \qquad = \frac{240}{30} \qquad = \frac{240}{40}$$

$$I_R = 9.6 \text{ A} \qquad I_Y = 8 \text{ A} \qquad I_B = 6 \text{ A}$$

The corresponding phasor diagrams are shown in Fig. 2.27.

Horizontal components, HC $= I_R - I_Y \cos 60° - I_B \cos 60°$

$$= 9.6 - (8 \times 0.5) - (6 \times 0.5)$$

so, HC $= 2.6$ A

Vertical components, VC $= I_B \sin 60° - I_Y \sin 60°$

$$= (6 \times 0.866) - (8 \times 0.866)$$

so VC $= -1.732$ A

The neutral current, $I_N = \sqrt{VC^2 + HC^2}$ amp

$$= \sqrt{-1.732^2 + 2.6^2}$$

hence, $I_N = 3.124$ A **Ans**

$$\phi = \tan^{-1}\frac{VC}{HC} = \tan^{-1}\frac{-1.732}{2.6}$$

hence, $\phi = -33.67°$ relative to I_R **Ans**

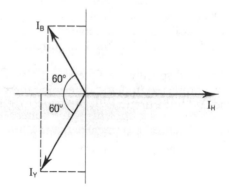

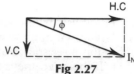

Fig 2.27

2.11 Advantages of Three-phase Systems

In previous sections, some of the advantages of three-phase systems, compared with single-phase systems, have been outlined. These advantages, together with others not yet described, are listed below. The advantages may be split into two distinct groups: those concerned with the generation and distribution of power, and those concerning a.c. motors. The last four of the advantages listed refer to the second group, and their significance will be appreciated when you have completed Chapter 5, which deals with a.c. machines.

1 The whole of the stator of a three-phase machine is utilised. A single phase alternator utilises only two-thirds of the stator slots.

2 A three-phase alternator has a better distribution factor.

3 A three-phase, four-wire system provides considerable savings in cable costs, for the distribution of an equivalent amount of power.

4 A three-phase, four-wire system provides alternative voltages for industrial and domestic users.

5 For a given machine frame size, the power output from a three-phase machine is greater than that from a single-phase machine.

6 A three-phase supply produces a rotating magnetic field; whereas a single-phase supply produces only a pulsating field.

7 Three-phase motors are inherently self-starting; whereas single-phase motors are not.

8 The torque produced by a three-phase motor is smooth; whereas that produced by a single-phase motor is pulsating.

Assignment Questions

1 A three-phase load is connected in star to a 400 V, 50 Hz supply. Each phase of the load consists of a coil, having inductance 0.2 H and resistance 40 Ω. Calculate the line current.

2 If the load specified in Question 1 above is now connected in delta, determine the values for phase and line currents.

3 A star-connected alternator stator generates 300 V in each of its stator windings. (a) Sketch the waveform and phasor diagrams for the phase voltages, and (b) calculate the p.d. between any pair of lines.

4 A balanced three-phase delta-connected load consists of the stator windings of an a.c. motor. Each winding has a resistance of 3.5 Ω and inductance 0.015 H. If this machine is connected to a 415 V, 50 Hz supply, calculate (a) the stator phase current, (b) the line current drawn from the supply, and (c) the total power dissipated.

5 Repeat the calculations for Question 4, when the stator windings are connected in star configuration.

6 Three identical coils, connected in star, take a total power of 1.8 kW at a power factor of 0.35, from a 415 V, 50 Hz supply. Determine the resistance and inductance of each coil.

7 Three inductors, each of resistance 12 Ω and inductance 0.02 H, are connected in delta to a 400 V, 50 Hz, three-phase supply. Calculate (a) the line current, (b) the power factor, and (c) the power consumed.

8 The star-connected secondary of a three-phase transformer supplies a delta-connected motor, which takes a power of 80 kW, at a lagging power factor of 0.85. If the line voltage is 600 V, calculate (a) the current in the transformer secondary windings, and (b) the current in the motor windings.

9 A star-connected load, each phase of which has an inductive reactance of 40 Ω and resistance of 25 Ω, is fed from the secondary of a three-phase, delta-connected transformer. If the transformer phase voltage is 600 V, calculate (a) the p.d. across each phase of the load, (b) the load phase current, (c) the current in the transformer secondary windings, and (d) the power and power factor.

10 Three coils are connected in delta to a 415 V, 50 Hz, three-phase supply, and take a line current of 4.8 A at a lagging power factor of 0.9. Determine (a) the resistance and inductance of each coil, and (b) the power consumed.

11 The power taken by a three-phase motor was measured using the two-wattmeter method. The readings were 850 W and 260 W respectively. Determine (a) the power consumption and, (b) the power factor of the motor.

12 Two wattmeters, connected to measure the power in a three-phase system, supplying a balanced load, indicate 10.6 kW and −2.4 kW respectively. Calculate (a) the total power consumed, and (b) the load phase angle and power factor. State the significance of the negative reading recorded.

13 Using the two-wattmeter method, the power in a three-phase system was measured. The meter readings were 120 W and 60 W respectively. Calculate (a) the power, and (b) the power factor.

14 Each branch of a three-phase, star-connected load, consists of a coil of resistance 4 Ω and reactance 5 Ω. This load is connected to a 400 V, 50 Hz supply. The power consumed is measured using the two-wattmeter method. Sketch a circuit diagram showing the wattmeter connections, and calculate the reading indicated by each meter.

15 A three-phase, 415 V, 50 Hz supply is connected to an unbalanced, star-connected load, having a power factor of 0.8 lagging in each phase. The currents are 40 A in the red phase, 55 A in the yellow phase, and 62 A in the blue phase. Determine (a) the value of the neutral current, and (b) the total power dissipated.

Suggested Practical Assignments

Assignment 1

To determine the relationship between line and phase quantities in three-phase systems.

Apparatus:
Low voltage, 50 Hz, three-phase supply
3 × 1 kΩ rheostats
3 × ammeters
3 × voltmeters (preferably DMM)

Method:

1 Using a digital meter, adjust each rheostat to exactly the same value (1 kΩ).

2 Connect the rheostats, in star configuration, to the three-phase supply.

3 Measure the three line voltages and the corresponding phase voltages, and record your results in a Table 1.

4 Measure the three line currents, and record these in Table 1.

5 Check that the neutral current is zero.

6 Carefully unbalance the load by altering the resistance value of one or two of the rheostats. ENSURE that you do not exceed the current ratings of the rheostats.

7 Measure and record the values of the three line currents and the neutral current, in a Table 2.

8 Switch off the three-phase supply, and disconnect the circuit.

9 Carefully reset the three rheostats to their original settings, as in paragraph 1 of the Method.

10 Connect the rheostats, in delta configuration, to the three-phase supply, with an ammeter in each line.

11 Switch on the supply, and measure the line voltages and line currents. Record your values in a Table 3.

12 Switch off the supply, and connect the three ammeters in the three phases of the load, i.e. an ammeter in series with each rheostat. This will involve opening each phase and inserting each ammeter.

13 Switch on and record the values of the three phase currents. Record values in Table 3.

14 From your tabulated readings, determine the relationship between line and phase quantities for both star and delta connections.

15 From your readings in Table 2, calculate the neutral current, and compare this result with the measured value.

16 Write an assignment report. Include all circuit and phasor diagrams, and calculations. State whether the line and phase relationships measured conform to those expected (allowing for experimental error).

Assignment 2

To measure the power in three-phase systems, using both single and two-wattmeter methods.

Apparatus:
Low voltage, 50 Hz, three-phase supply
3×1 kΩ rheostats
$2 \times$ wattmeters
$1 \times$ DMM

Method:

1 As for Assignment 1, carefully adjust the three rheostats to the same value (1 kΩ).

2 Connect the circuit shown in Fig. 2.28, and measure the power in the red and yellow phases.

3 Transfer one of the wattmeters to the blue phase, and measure that power.

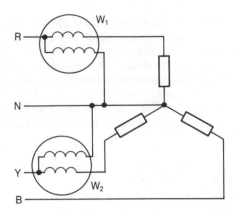

Fig 2.28

4 Add the three wattmeter readings to give the total power dissipation. Check to see whether this is three times the individual phase power.

5 Switch off the power supply, and reconnect as in Fig. 2.29.

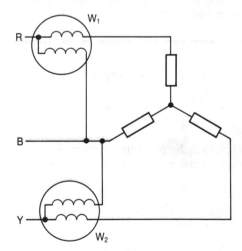

Fig 2.29

6 Record the two wattmeter readings, and check whether their sum is equal to the total power recorded from paragraph 4 above.

7 Switch off the power supply and reconnect the circuit as in Fig. 2.30.

8 Record the two wattmeter readings.

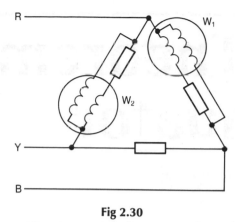

Fig 2.30

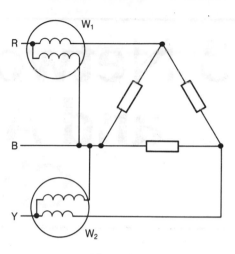

Fig 2.31

9 Switch off the supply, and transfer one of the wattmeters to the third phase.

10 Record this reading. Add the three readings to give the total power, and check that this is three times the phase power.

11 Switch off the supply and connect the circuit of Fig. 2.31.

12 Record the two wattmeter readings, and check that their sum equals the total power obtained from paragraph 10 above.

Assignment 3

Show how the two-wattmeter method of power measurement can be accomplished by means of a single wattmeter, together with a suitable switching arrangement. You may either devise your own system, or discover an existing system through library research.

▪ 3 Network Theorems and Attenuator Circuits ▪

This chapter introduces a number of network theorems, and their application to the solution of network problems. The decibel notation and its usage, together with simple attenuator circuits is also covered. On completion you should be able to:

1] Apply the principle of the Superposition Theorem to the solution of circuits.

2] Explain and use the concepts of constant voltage and constant current sources, and convert from one to the other.

3] Solve circuit problems by means of Thévénin's and Norton's Theorems.

4] Apply the Maximum Power Transfer Theorem, and transformer matching technique.

5] Define and use decibel notation for power, voltage and current gain/attenuation.

6] Describe the function of simple attenuator networks, and solve problems related to these.

3.1 Superposition Theorem

This theorem provides an alternative to the application of Kirchhoff's laws to the solution of networks containing more than one source of emf. The Superposition principle allows us to break up a complex circuit into separate sections, and then apply simple Ohm's law techniques to each section in turn. The theorem is defined as follows:

In any network consisting of resistors, the current flowing in any branch is the algebraic sum of the currents that would flow in that branch, if each emf source was considered separately, all others being replaced by resistors equal to their internal resistance.

In the above definition the term resistance has been used. This is because you are required to deal only with resistive circuits. However, the theorem applies equally to a.c. circuits, which may also include reactive elements. In this case, the word *impedance* should be substituted for resistance. This comment applies also to the other network theorems, that follow shortly.

As with most theorems, it sounds most complicated. However, putting the principle into practice is relatively simple, and is best illustrated by means of an example.

■ Worked Example 3.1 ■

Q A circuit containing two sources of emf is shown in Fig. 3.1. Using the principle of Superposition, determine the current flowing in the 5 Ω resistor.

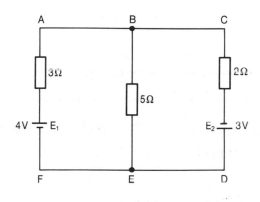

Fig 3.1

A

The theorem allows us to split the original circuit into two sub-circuits (shown in Figs. 3.2 and 3.3), each of

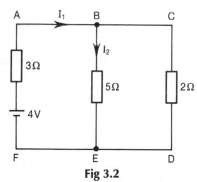

Fig 3.2

which contains only one battery, the other battery being replaced by its internal resistance. Note that it is not clear, from the diagram, whether the 3 Ω and 2 Ω resistors are the internal resistances of E_1 and E_2 respectively; or that E_1 and E_2 are ideal sources (having

zero internal resistance), with the two resistors being external resistors in the circuit. In this case it makes no difference, so the two sub-circuits shown apply in both circumstances.

Considering Fig. 3.2, currents I_1 and I_2 may be calculated as follows:

Total resistance of the circuit, R consists of the 3 Ω in series with the parallel combination of the 5 Ω and the 2 Ω.

$$\text{Hence, } R = 3 + \frac{2 \times 5}{2 + 5} = 4.429 \ \Omega$$

$$I_1 = \frac{E_1}{R} \text{ amp} = \frac{4}{4.429}$$

$$\text{so, } I_1 = 0.903 \text{ A}$$

Using current divider technique, current I_2 is:

$$I_2 = \frac{2}{7} \times 0.903 = 0.258 \text{ A from B to E} \ \dots\dots\dots \ [1]$$

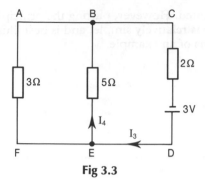

Fig 3.3

Considering Fig. 3.3, we can apply the same techniques to evaluate currents I_3 and I_4:

$$R = 2 + \frac{5 \times 3}{5 + 3} = 3.875 \ \Omega$$

$$I_3 = \frac{3}{3.875} = 0.774 \text{ A}$$

$$I_4 = \frac{3}{8} \times 0.774 = 0.29 \text{ A from E to B} \ \text{.......} \ [2]$$

The current through the 5 Ω resistor is the algebraic sum of the currents I_2 and I_4 as in [1] and [2] above. However, notice that these two currents are in opposite directions. The algebraic sum is therefore the *difference* between them.

Hence, current through the 5 Ω = $I_4 - I_2$ amp

= 0.032 A from E to B **Ans**

3.2 Constant Voltage and Constant Current Sources and their Equivalence

Thus far, we have considered only voltage sources, in the form of a constant emf in series with its internal resistance. In some circuits it is more convenient to consider the source as a constant current generator, with its internal resistance in parallel with it. This is often the case when dealing with the equivalent circuit for a transistor.

Consider a constant voltage source of emf E volt and internal resistance r ohm, supplying a load resistor R, as shown in Fig. 3.4.

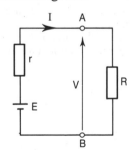

Fig 3.4

$$\text{load current}, \ I = \frac{E}{r + R} \text{ amp} \ \text{....} \ [1]$$

$$\text{and terminal p.d.}, \quad V = IR \text{ volt} \ \text{.........} \ [2]$$

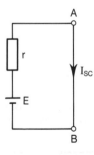

Fig 3.5

Let the load R now be replaced by a short-circuit between terminals A and B, as shown in Fig. 3.5.

$$I_{sc} = \frac{E}{r} \text{ amp} \ \text{.........} \ [3]$$

Let us now replace the original voltage source by a constant current source, of value I_{sc} and internal resistance of r ohm in parallel with it. This is shown in Fig. 3.6. The linked circles form the circuit symbol for a current generator.

Considering this figure, the current through the load, and the terminal p.d. may be determined by applying the current divider technique as follows:

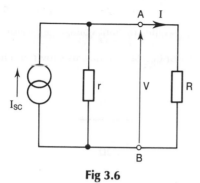

Fig 3.6

$$I = \frac{r}{r+R} \times I_{sc} \text{ amp}$$

and substituting [3] for I_{sc}:

$$I = \frac{r}{r+R} \times \frac{E}{r}$$

so $I = \dfrac{E}{r+R}$ amp .... [4]

and $V = IR$ volt .... [5]

Since the expressions for the load current and terminal voltage in [1] and [2] are identical to those in [4] and [5] respectively, then it follows that the constant voltage generator of Fig. 3.4 is equivalent to the constant current generator of Fig. 3.6, and vice-versa.

▪ **Worked Example 3.2** ▪

Q (a) Convert the voltage source shown in Fig. 3.7 into its equivalent constant current source, and (b) convert the current source of Fig. 3.8 into its equivalent voltage source.

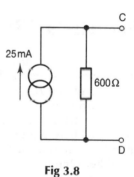

Fig 3.8

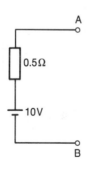

Fig 3.7

A ────────────────────────────

(a) Placing a short-circuit between the terminals of the voltage generator will produce a current of:

$$I_{sc} = \frac{10}{0.5} \text{ amp} = 20 \text{ A}$$

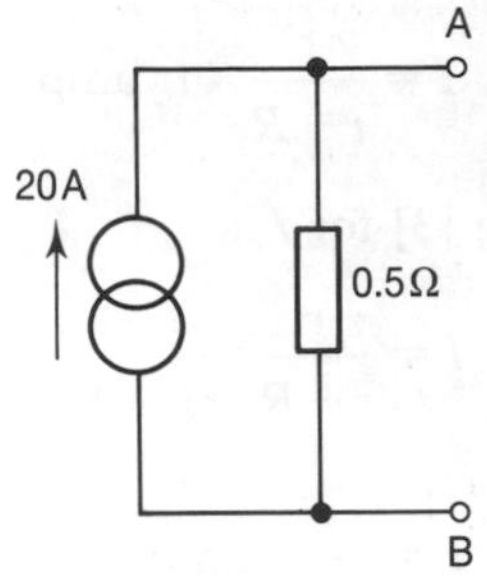

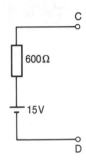

Fig 3.9

so the constant current generator will produce a current of 20 A, with an internal resistance of 0.5 Ω in parallel, as shown in Fig. 3.9.

(b) From Fig. 3.8, the open circuit terminal voltage is:

$$V = 25 \times 10^{-3} \times 600 = 15 \text{ V}$$

so the constant voltage generator will have an emf of 15 V, with an internal resistance of 600 Ω in series, as shown in Fig. 3.10.

Fig 3.10

3.3 Thévénin's Theorem

Thévénin's theorem states that any network containing sources and resistors can be replaced by a *single* voltage source of emf E_o and series internal resistance R_o, as shown in Fig. 3.11. The value of E_o is equal to the open-circuit terminal voltage of the network, measured between terminals A and B. The resistance R_o is the resistance of the network measured between terminals A and B, with all network sources replaced by their internal resistances.

Note: If any of the sources in the original network are shown as being ideal (zero internal resistance), these will be replaced by a short-circuit.

The application of Thévénin's theorem to the solution of network is illustrated in the following example.

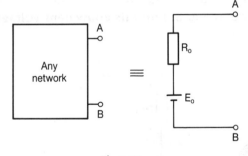

Fig 3.11

■ Worked Example 3.3 ■

Q For the network shown in Fig. 3.12, (a) determine the Thévenin equivalent generator for the circuit to the left of terminals A and B, (b) hence calculate the p.d. across and the current flowing through the 5 Ω load resistor, and (c) calculate the p.d. across and the current through the load if this resistance is changed to 3 Ω.

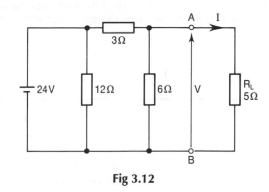

Fig 3.12

A

(a) With the 5 Ω resistor removed, terminals A and B will be open-circuit. The full 24 V will be developed across the 12 Ω resistor, so the 6 Ω and 3 Ω resistors form a simple potential divider.

$$\text{Thus, } E_o = \frac{6}{9} \times 24 = 16 \text{ V}$$

Since the battery is shown as an ideal source, then it is replaced by a short-circuit. Thus, 'looking in' at the terminals, the circuit appears as in Fig. 3.13. As the

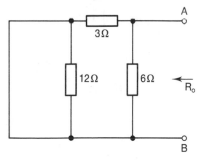

Fig 3.13

12 Ω resistor is now short-circuited, then the 6 Ω and the 3 Ω resistors are effectively connected in parallel.

$$\text{Hence, } R_o = \frac{3 \times 6}{9} = 2 \text{ Ω}$$

The Thévenin equivalent generator therefore has an emf of 16 V and internal resistance of 2 Ω. The complete equivalent circuit is shown in Fig. 3.14.

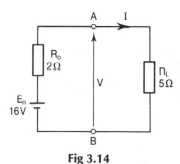

Fig 3.14

(b) From Fig. 3.14:

$$V = \frac{5}{7} \times 16 = 11.43 \text{ V } \textbf{Ans}$$

$$\text{and } I = \frac{V}{R_L} \text{ amp} = \frac{11.43}{5}$$

$$\text{so, } I = 2.29 \text{ A } \textbf{Ans}$$

(c)

$$V = \frac{3}{5} \times 16 = 9.6 \text{ V } \textbf{Ans}$$

$$\text{and } I = \frac{9.6}{3} = 3.2 \text{ A } \textbf{Ans}$$

Note: The original circuit could have been solved using normal Ohm's law techniques, but the calculations would have been more extensive. The real advantage of deriving the Thévénin equivalent generator is that the load value may be changed to any value, as many times as is wished, but the subsequent calculations for *V* and *I* would be very quick and simple. Using normal Ohm's law techniques, *all* the circuit currents etc. would have to be recalculated each time.

■ **Worked Example 3.4** ■

Q For the circuit of Fig. 3.15,
(a) determine the Thévénin equivalent generator for the network to the left of terminals A and B, and hence calculate the current flowing through the 5 Ω load resistor, and (b) determine the load current if the load resistor is changed to 10 Ω.

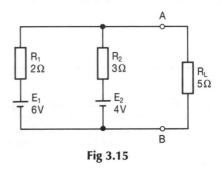

Fig 3.15

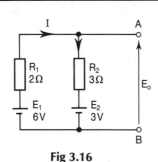

Fig 3.16

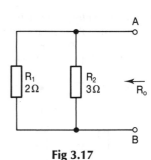

Fig 3.17

(a) Removing the load resistor, the circuit is as shown in Fig. 3.16, and the terminal p.d. (E_o) calculated as follows.

$$I = \frac{E_1 - E_2}{R_1 + R_2} \text{ amp} = \frac{6 - 4}{2 + 3}$$

therefore, $I = 0.4$ A

$$E_o = E_1 - IR_1 \quad \text{or } E_2 + IR_2 \text{ volt}$$
$$= 6 - (0.4 \times 2) \quad \text{or } 4 + (0.4 \times 3)$$

so, $E_o = 5.2$ V

From Fig. 3.17, where the two sources have been replaced by their internal resistances:

$$R_o = \frac{R_1 R_2}{R_1 + R_2} \text{ ohm} = \frac{6}{5} = 1.2 \; \Omega$$

Hence, the Thévénin equivalent circuit will be as shown in Fig. 3.18, and from this circuit:

$$I = \frac{E_o}{R_o + R_L} \text{ amp} = \frac{5.2}{6.2}$$

so, $I = 0.839$ A **Ans**

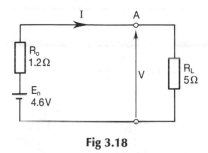

Fig 3.18

(b) With R_L changed to 10 Ω, and using the Thévénin circuit, the load current is simply obtained thus:

$$I = \frac{E_o}{R_o + R_L} \text{ amp} = \frac{5.2}{11.2} = 0.464 \text{ A } \textbf{Ans}$$

Note: This problem could have been solved using Kirchhoff's laws (or Superposition theorem). However, in order to complete part (b), all of the loop equations would have to be derived again, to provide a new set of simultaneous equations for solution. This clearly illustrates the advantage of using the Thévénin generator.

3.4 Norton's Theorem

This theorem complements Thévénin's in that the network is replaced by a constant current generator of I_{sc} amp, and parallel internal resistance R_o ohm. The current, I_{sc}, is the current that would flow between the terminals when they are short-circuited. The resistance R_o is defined in exactly the same way as it is for the Thévénin circuit. Thus R_o will be the same for both a Norton and a Thévénin equivalent circuit. The Norton constant current generator is illustrated in Fig. 3.19.

The application of Norton's theorem is illustrated in the following example.

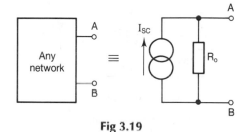

Fig 3.19

■ Worked Example 3.5 ■

Q Derive the Norton equivalent circuit for the network to the left of terminals A and B in Fig. 3.20, and hence determine the current through the 10 Ω load resistor.

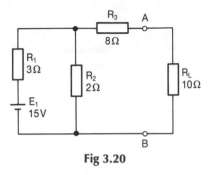

Fig 3.20

A

With terminals A and B short-circuited, the circuit will

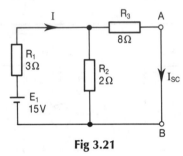

Fig 3.21

be as in Fig. 3.21. Using this circuit, the short-circuit current I_{sc} is found as follows:

$$\text{Total resistance, } R = R_1 + \frac{R_2 R_3}{R_2 + R_3} \text{ ohm} = 3 + \frac{16}{10}$$

$$\text{so, } R = 4.6\ \Omega$$

$$I = \frac{E_1}{R} \text{ amp} = \frac{15}{4.6} = 3.261\ A$$

Using current division,

$$I_{sc} = \frac{R_2}{R_2 + R_3} \times I \text{ amp} = \frac{2}{10} \times 3.261$$

$$\text{therefore, } I_{sc} = 0.652\ A$$

From Fig. 3.22, the resistance R_o is obtained thus:

$$R_o = R_3 + \frac{R_1 R_2}{R_1 + R_2} \text{ ohm} = 8 + \frac{6}{5}$$

$$\text{so, } R_o = 9.2\ \Omega$$

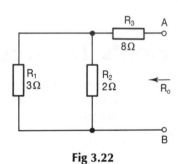

Fig 3.22

The complete Norton equivalent circuit will be as shown in Fig. 3.23, and from this figure:

$$I = \frac{R_o}{R_o + R_L} \times I_{sc} \text{ amp} = \frac{9.2}{19.2} \times 0.652$$

$$\text{hence, } I = 0.312\ A \text{ **Ans**}$$

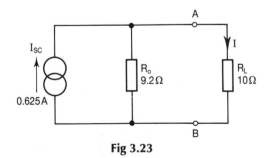

Fig 3.23

The advantage of the Norton equivalent circuit is the same as that for the Thévénin circuit; i.e. if the load resistance value is repeatedly changed, the new load current can be very simply and rapidly calculated.

3.5 The Maximum Power Transfer Theorem

This theorem states that maximum power is transferred from a source to a load when the load resistance is of the same value as the internal resistance of the source.

Again, note that this theorem also applies to a.c. circuits; in which case the word *resistance* should be replaced by the word *impedance*. However, the generator may also possess internal reactance. In this case, for maximum power transfer, the load must also possess an equal value of reactance, but of the opposite type. In other words, if the source has inductive reactance, then the load must have an equivalent capacitive reactance; and vice-versa.

■ Worked Example 3.6 ■

  Q For the circuit shown in Fig. 3.20 (Example 3.5), determine the value of load resistor R_L that would result in maximum power dissipation in this load, and calculate the value of this power.

A ———————————————————————————

In order to solve this problem, the Norton (or Thévénin) equivalent circuit would be derived, as in Example 3.5. Thus, for this example, we know that the equivalent Norton generator has a current of 0.652 A, and internal resistance of 9.2 Ω. Hence the value of R_L that will result in maximum power transfer will be 9.2 Ω **Ans**.

The generator current, I_{sc} will divide equally between R_o and R_L, hence:

$$\text{the load current, } I = \frac{0.652}{2} = 0.326 \text{ A}$$

$$P_L = I^2 R_L \text{ watt} = 0.326^2 \times 9.2$$

$$\text{therefore } P_L = 0.978 \text{ W } \textbf{Ans}$$

In order to ensure that maximum power is transferred from a source to a load, the load has to be matched to the load. This effect can be achieved in a.c. circuits by means of an impedance matching transformer. Using this technique, the source is effectively connected to a load of impedance value equal to its own internal resistance. To illustrate the principle, consider a single-phase transformer, connected to a resistive load, as shown in Fig. 3.24.

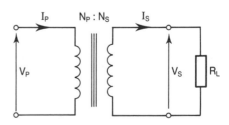

Fig 3.24

The primary and secondary windings will be designed to have only a very small resistance (i.e. we may consider them to be perfect inductors). Thus the power developed in the secondary circuit must be due only to the secondary current flowing through the load resistor, R_L. However, the secondary of the transformer must derive its power from the primary circuit. The power in the primary circuit must therefore be due to the primary current flowing through some *effective* resistance. This effective resistance is known as 'the secondary resistance, referred to the primary'. In other words, the supply connected to the primary winding 'sees' an effective resistance connected between the primary terminals. The value of this referred resistance is obtained as follows.

$$R_L = \frac{V_s}{I_s} \text{ ohm; and } R_p = \frac{V_p}{I_p} \text{ ohm}$$

so, $\dfrac{R_p}{R_L} = \dfrac{V_p}{I_p} \times \dfrac{I_s}{V_s} = \dfrac{V_p}{V_s} \times \dfrac{I_s}{I_p}$

but, $\dfrac{V_p}{V_s} = \dfrac{N_p}{N_s}$ and $\dfrac{I_s}{I_p} = \dfrac{N_p}{N_s}$

therefore, $\dfrac{R_p}{R_L} = \left(\dfrac{N_p}{N_s}\right)^2$

hence, $R_p = \left(\dfrac{N_p}{N_s}\right)^2 \times R_L$ (3.1)

i.e. The resistance referred to the primary circuit is equal to the load resistance multiplied by the *square* of the turns ratio.

■ Worked Example 3.7 ■

Q A loudspeaker of impedance 8 Ω is fed from an amplifier of output impedance 1.8 k Ω. Determine the turns ratio of a suitable matching transformer.

A ───────────────────────────

$R_p = 1800\ \Omega;\ R_L = 8\ \Omega$

$$R_p = \left(\frac{N_p}{N_s}\right)^2 R_L \text{ ohm}$$

so, $\left(\dfrac{N_p}{N_s}\right)^2 = \dfrac{R_p}{R_L} = \dfrac{1800}{8}$

therefore, $\left(\dfrac{N_p}{N_s}\right)^2 = 225$

hence, $\dfrac{N_p}{N_s} = 15$ i.e. turns ratio of 15 : 1 **Ans**

3.6 The Decibel and its Usage

Consider a two-stage amplifier system, as shown in Fig. 3.25. Each amplifier provides an increase of

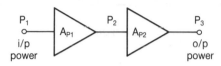

Fig 3.25

the *signal* power. This effect is referred to as the power gain, A_p, of the amplifier. This means that the signal output power from an amplifier is greater than its signal input power. This power gain may be expressed as:

$$A_p = \frac{P_{out}}{P_{in}} \quad \text{or} \quad \frac{P_o}{P_i}$$

For the system shown, if the input power P_1 is 1 mW, and the gains of the two amplifier stages are 10 times and 100 times respectively, then the final output power, P_3, may be determined as follows.

$$A_{p1} = \frac{P_2}{P_1}; \quad \text{so} \quad P_2 = A_{p1} \times P_1 \text{ watt}$$

therefore, $P_2 = 10 \times 10^{-3} = 10 \text{ mW}$

$$A_{p2} = \frac{P_3}{P_2}; \quad \text{so} \quad P_3 = A_{p2} \times P_2 \text{ watt}$$

therefore, $P_3 = 100 \times 10 \times 10^{-3} = 1 \text{ W}$

From these results it may be seen that the output power of the system is 1000 times that of the input.

In other words, overall signal power gain,

$$A_p = \frac{P_3}{P_1} = \frac{1}{10^{-3}} = 1000$$

or, overall signal power gain,

$$A_p = A_{p1} A_{p2}$$

In general, when amplifiers (or other devices) are cascaded in this way, the overall gain (or loss) is given by the product of the individual stage gains (or losses).

Note: The efficiency of any machine or device is defined as the ratio of its output power to its input power. However, this does NOT mean that an amplifier is more than 100% efficient! The reason is that only the *signal* input and output powers are considered when quoting the power gain. No account is taken of the comparatively large amount of power injected from the d.c. power supply, without which the amplifier cannot function. In practice, small signal voltage amplifiers will have an *efficiency* figure of less than 25%. Power amplifiers may have an efficiency in the order of 70%.

It is often more convenient to express power gain ratios in a logarithmic form, known as the Bel (named after Alexander Graham Bell). Thus a power gain expressed in this way is:

$$A_p = \log \frac{P_o}{P_i} \text{ Bel}$$

where P_i = input power; P_o = output power; and logarithms to the base 10 are used.

The Bel is an inconveniently large unit for practical purposes, so the decibel (one tenth of a Bel) is used.

$$\text{Hence, } A_p = 10 \log \frac{P_o}{P_i} \text{ decibel} \tag{3.1}$$

The unit symbol for the decibel is dB. For the two-stage amplifier system considered, the power gains would be expressed as follows:

$$A_{p1} = 10 \log 10 = 10 \text{ dB}$$

$$A_{p2} = 10 \log 100 = 20 \text{ dB}$$

$$\text{and } A_p = 10 \log 1000 = 30 \text{ dB}$$

Note that the overall system gain, A_p, when expressed in dB is simply the sum of the individual stage gains, also expressed in dB.

▪ Worked Example 3.8 ▪

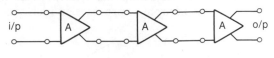

Fig 3.26

Q A communications system, involving transmission lines and amplifiers, is illustrated in Fig. 3.26. Each section of transmission line attenuates (reduces) the signal power by a factor of 35.5%, and each amplifier has a gain ratio of 5 times. Calculate the overall power gain of the system as (a) a power ratio, and (b) in decibels.

A _____

(a) For each line,
$$\frac{P_o}{P_i} = 0.355$$

so, total loss $= 0.355 \times 0.355 \times 0.355$
$$= 0.0159$$

For each amplifier, $\dfrac{P_o}{P_i} = 5$

so, total gain $= 5 \times 5 \times 5 = 125$

therefore, overall gain $= 125 \times 0.0159$
$$= 2 \text{ times } \textbf{Ans}$$

(b) For each line, attenuation is
$$10 \log 0.355 = -4.5 \text{ dB*}$$
so total attenuation $= -4.5 \times 4$
$$= -18 \text{ dB}$$

For each amplifier, gain $= 10 \log 5$
$$= +7 \text{ dB}$$
so total gain $= 7 \times 3 = 21 \text{ dB}$

Hence, overall gain of the system $= 21 - 18 = 3$ dB **Ans**

*Note that a loss or attenuation expressed in dB has a negative value; whereas a gain has a positive value. A further point to note is that if the gains and attenuations of the system had originally been expressed in dB, then the calculation would simply have been as follows:

Overall system gain $= (3 \times 7) - (4 \times 4.5) = 3$ dB **Ans**

The above example illustrates the convenience of using decibel notation, since it involves only simple addition and subtraction to determine the overall gain or attenuation of a system. It has also been shown that a gain of 2 times is equivalent to a power gain of 3 dB (more precisely, 3.01 dB). It is left to the reader to confirm, by using a calculator, that an attenuation of 2 times (i.e. a gain of 0.5) is equal to -3 dB. This figure of -3 dB will frequently be met when dealing with the frequency response curves for amplifiers and other frequency-dependent circuits, such as series and parallel tuned circuits.

In order to gain a 'feel' for power gains and losses expressed in dB, the following list shows the corresponding power gain ratios.

$$
\text{gains}
\begin{cases}
10\,000 \text{ times} = 40 \text{ dB} \\
1\,000 \text{ times} = 30 \text{ dB} \\
100 \text{ times} = 20 \text{ dB} \\
10 \text{ times} = 10 \text{ dB} \\
2 \text{ times} = 3 \text{ dB} \\
1 \text{ times} = 0 \text{ dB}
\end{cases}
$$

$$
\text{losses}
\begin{cases}
0.5 \text{ times} = -3 \text{ dB} \\
0.1 \text{ times} = -10 \text{ dB} \\
0.01 \text{ times} = -20 \text{ dB} \\
0.001 \text{ times} = -30 \text{ dB} \\
0.0001 \text{ times} = -40 \text{ dB}
\end{cases}
$$

▪ Worked Example 3.9 ▪

(a) Convert the following gain ratios into dB.

(i) 250, (ii) 50, (iii) 0.4

(b) Convert the following gains and losses into ratios.

(i) 25 dB, (ii) 8 dB, (iii) −15 dB

(a)

(i) $A_p = 10 \log 250 = 24$ dB **Ans**

(ii) $A_p = 10 \log 50 = 17$ dB **Ans**

(iii) $A_p = 10 \log 0.4 = -4$ dB **Ans**

(b)

(i) $25 = 10 \log$ (ratio) dB

so $2.5 = \log$ (ratio)

therefore, (ratio) = antilog $2.5 = 316$ times **Ans**

(ii) $8 = 10 \log$ (ratio) dB

$0.8 = \log$ (ratio)

therefore, (ratio) = antilog $0.8 = 6.3$ times **Ans**

(iii) $-15 = 10 \log$ (ratio) dB

$-1.5 = \log$ (ratio)

therefore, (ratio) = antilog $(-1.5) = 0.032$ **Ans**

Although the decibel is defined in terms of a power ratio, it may also be used to express both voltage and current ratios, provided that certain conditions are met. These conditions are that the resistance of the load is the same as that of the source, i.e. the conditions for maximum power transfer. Consider such a system, whereby the two resistance values are R ohm, the input voltage is V_1 volt, and the output voltage is V_2 volt. Let the corresponding currents be I_1 and I_2 ampere.

$$P_1 = \frac{V_1^{\,2}}{R} \text{ watt and } P_2 = \frac{V_2^{\,2}}{R} \text{ watt}$$

$$\text{gain} = 10 \log\left(\frac{V_2^{\,2}}{R} \times \frac{R}{V_1^{\,2}}\right)$$

$$= 10 \log\left(\frac{V_2}{V_1}\right)^2$$

hence, voltage gain,

$$A_v = 20 \log \frac{V_2}{V_1} \text{ dB} \qquad (3.2)$$

Also, using the fact that $P_1 = I_1^{\,2}R$ watt and $P_2 = I_2^{\,2}R$ watt it is left to the reader to verify that:

$$\text{current gain, } A_i = 20 \log \frac{I_2}{I_1} \text{ dB} \qquad (3.3)$$

It was shown in the previous chapter, that a tuned circuit used as a pass-band or stop-band filter has a bandwidth. The same concept also applies to a.c. amplifiers. In the latter case, the frequency response curve for a voltage amplifier would be similar to that shown in Fig. 3.27. The bandwidth is defined as that range of frequencies over which the voltage gain is greater than, or equal to $A_{vm}/\sqrt{2}$, where A_{vm} is the mid-frequency gain. A similar response curve for the amplifier current gain could also be plotted.

Now, power gain = voltage gain × current gain

or, $A_p = A_v \times A_i$

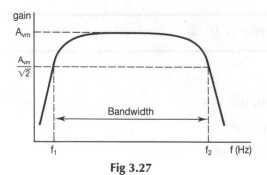

Fig 3.27

$$A_v = \frac{A_{vm}}{\sqrt{2}} \quad \text{and} \quad A_i = \frac{A_{im}}{\sqrt{2}} \text{ respectively}$$

$$\text{thus } A_p = \frac{A_{vm}}{\sqrt{2}} \times \frac{A_{im}}{\sqrt{2}} = \frac{A_{pm}}{2}$$

These points on the response curve are therefore referred to as either the cut-off points, the half-power points, or the −3 dB points.

The cut-off frequencies, f_1 and f_2, define the bandwidth, and at these frequencies, the current and voltage gains will be:

▪ **Worked Example 3.10** ▪

Q An amplifier is fed with a 50 mV, 200 μA signal. The amplifier has a voltage gain of 75 times and a current gain of 150 times. Determine (a) the voltage, current and power gains, expressed in dB, and (b) the output voltage, current and power. You may assume that input and output resistances are the same value.

A
_____

$V_1 = 50 \times 10^{-3}$ V; $I_1 = 200 \times 10^{-6}$ A; $A_v = 75$; $A_i = 150$

    (a) $A_v = 20 \log 75$ dB $= 37.5$ dB **Ans**

                 $A_i = 20 \log 150$ dB $= 43.5$ dB **Ans**

                 $A_p = A_v \times A_i = 75 \times 150 = 11\,250$ times

        so, $A_p = 10 \log 11\,250$ dB $= 40.5$ dB **Ans**

    (b) $V_2 = V_1 \times A_v$ volt $= 50 \times 10^{-3} \times 75 = 3.75$ V **Ans**

                 $I_2 = I_1 \times A_i$ amp $= 200 \times 10^{-6} \times 150 = 30$ mA **Ans**

                 $P_2 = V_2 I_2$ watt $= 3.75 \times 0.03 = 112.5$ mW **Ans**

▪ Worked Example 3.11 ▪

$\boxed{Q}$ An amplifier has a voltage gain of 10 times at a frequency of 150 Hz; 60 times between 2 kHz and 12 kHz; and 15 times at 35 kHz. Determine (a) the voltage gain, expressed in decibel, for each case, and (b) the voltage gain at the limits of its bandwidth.

$\boxed{A}$ ──

(a) at 150 Hz: $A_v = 20 \log 10 = 20$ dB **Ans**

at mid-frequencies: $A_v = 20 \log 60 = 35.6$ dB **Ans**

at 35 kHz: $A_v = 20 \log 15 = 23.5$ dB **Ans**

(b) The limits of the bandwidth occur when A_v is 3 dB down on the mid-frequency value, or $A_v = A_{vm}/\sqrt{2}$.

Therefore, $A_v = A_{vm} - 3 = 35.6 - 3 = 32.6$ dB **Ans**

or, $A_v = \dfrac{A_{vm}}{\sqrt{2}} = \dfrac{60}{\sqrt{2}} = 42.426$ times

hence, $A_v = 20 \log 42.426 = 32.6$ dB **Ans**

The decibel is defined in terms of a power *ratio*. Thus, the power output of a device cannot be quoted directly in dB. For example, to say that an amplifier has an output power of 20 dB is completely meaningless. In this case it only makes sense to say that the amplifier has a power *gain* of 20 dB. However, actual values of output power may be expressed in decibel form *provided* a known reference power input is used. In electronic work, this reference power is one milliwatt (1 mW), and powers expressed in this way are designated as dBm.

Considering such an amplifier, it could be said to have a power output of 20 dBm. This would mean that its actual output power was 0.1 W. This

figure is obtained thus:

$$10 \log \frac{P_2}{P_1} = 20 \text{ dBm}$$

$$\log \frac{P_2}{P_1} = 2$$

$$\frac{P_2}{P_1} = \text{antilog } 2 = 100$$

and since $P_1 = 1$ mW, then P_2 $100 \times 10^{-3} = 0.1$ W or 100 mW.

3.7 Attenuator Networks ──────────────

An attenuator is a network designed to provide a predetermined voltage reduction (attenuation) between its input and output terminals. Attenuators are usually resistor networks, because resistance is independent of frequency. This avoids the problem of the selected attenuation varying as the frequency varies. Another feature is the

desirability of maximum power transfer from the original source to the load. In this latter case, the attenuator would be designed so that its input resistance (impedance) is equal to the source resistance (impedance). Consider the following two examples.

■ Worked Example 3.12 ■

 The circuit of Fig. 3.28 shows a simple resistive attenuator network, supplied at 10 V (V_i). Its output terminals are connected to a 75 Ω load. Calculate (a) the output voltage (V_o), and (b) the resulting attenuation, expressed both as a voltage ratio, and in decibels.

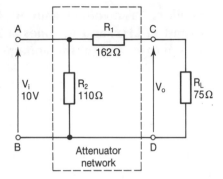

Fig 3.28

A ────────────────────────────────

In this circuit, R_1 and R_L form a simple potential divider, supplied at 10 V.

(a)
$$\text{Thus, } V_o = \frac{R_L}{R_1 + R_L} \times V_i = \frac{75}{237} \times 10$$

hence, $V_o = 3.165$ V **Ans**

(b)
$$\frac{3.165}{10} = 0.3165 \text{ **Ans**}$$

and expressing this in decibels:

$$\frac{V_o}{V_i} = 20 \log 0.3165 \text{ dB}$$

so, attenuation = −10 dB **Ans**

It should be noted that the attenuator circuit in the above example provides a partial match between the source and the load. By this it is meant that the resistance 'seen' by the source between terminals A and B is the effective resistance of R_2 in parallel with the series combination of R_1 and R_L. A simple calculation shows this resistance value to be 75 Ω.

Thus, the attenuator imposes exactly the same loading effect on the source as the load resistor would, if the attenuator was not there. However, this will not provide maximum power transfer. To achieve the latter effect, consider the next example.

▪ **Worked Example 3.13** ▪

Q A source of emf 10 V and internal resistance 600 Ω is connected to a 600 Ω load via a resistive attenuator network as shown in Fig. 3.29. Calculate (a) the output voltage (V_o), (b) the attenuator input voltage (V_i), and (c) the attenuation expressed in decibels.

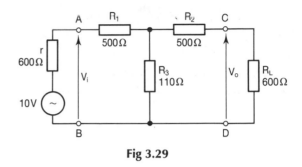

Fig 3.29

A

(a) Firstly, let us determine the Thévenin equivalent circuit to the left of terminals C and D.

$$R_{CD} = \frac{(r + R_1)R_3}{(r + R_1) + R_3} + R_2 \text{ ohm}$$

$$= \frac{(600 + 500) \times 110}{600 + 500 + 110} + 500$$

so, $R_{CD} = 100 + 500 = 600 \ \Omega$

Open-circuit p.d. across CD = open-circuit p.d. across R_3

therefore, $V_{CD} = \dfrac{R_3}{r + R_1 + R_3} \times 10$ volt

(simple potential divider)

so, $V_{CD} = \dfrac{110}{1210} \times 10 = 0.9091$ V

The Thévenin equivalent circuit is shown in Fig. 3.30, and from this we can say:

$$V_o = \frac{600}{1200} \times 0.9091 = 0.4545 \text{ V } \textbf{Ans}$$

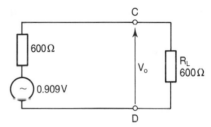

Fig 3.30

(b) Since the whole circuit is symmetrical, then the resistance looking into terminals AB will be the same as that looking back into terminals CD, i.e. $R_{AB} = 600 \ \Omega$.

Hence, $V_i = \dfrac{R_{AB}}{r + R_{AB}} \times 10$ volt $= \dfrac{600}{1200} \times 10$

so $V_i = 5$ V **Ans**

(c)

$$\text{Attenuation} = 20 \log \frac{V_o}{V_i} \text{ dB} = 20 \log \frac{0.4545}{5}$$

so attenuation $= -21$ dB **Ans**

The attenuator network in the last example matches both the source and load resistances.

Thus, maximum power will be transferred from source to load.

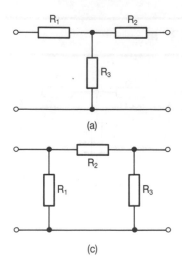

(a)

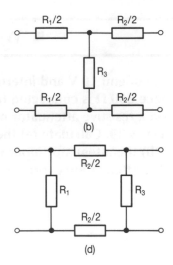

(b)

(c)

(d)

Fig 3.31

Both of the previous two examples were concerned with a fixed value of attenuation. An attenuator network of this type is often referred to as a *pad*. Attenuators usually take the form of either a 'T' or a 'π' network, and they may be balanced or unbalanced; symmetrical or asymmetrical. The differences between these types of pad are illustrated in Figs. 3.31(a) to (d).

Fig. 3.31(a) shows an unbalanced 'T' pad. It will also be symmetrical if $R_1 = R_2$.

Fig. 3.31(b) shows a balanced 'T' pad. It will also be symmetrical if $R_1/2 = R_2/2$.

Fig. 3.31(c) shows an unbalanced 'π' pad. It will be symmetrical if $R_1 = R_3$.

Fig. 3.31(d) shows a balanced 'π' pad. This will also be symmetrical if $R_1 = R_3$.

In addition to the pads described above, it is also possible to have variable attenuators. These may consist of a series of symmetrical pads that can be cascaded (connected in series) by suitable switching. An example of this is a decade step attenuator. This device contains the pads inside a metal case, with say three switches, plus the two input and two output terminals. The switches allow the pads to be interconnected so as to select attenuation figures in steps of 10 dB, 1 dB and 0.1 dB. If, for example, the switches were set to select 20 dB, 5 dB and 0.3 dB respectively, the overall attenuation selected would be 25.3 dB. Remember, when using decibel notation, we *add* the individual attenuation figures to give the total value.

Assignment Questions

1 Using the principle of Superposition, determine the value of current flowing through the 6 Ω resistor of the circuit shown in Fig. 3.32.

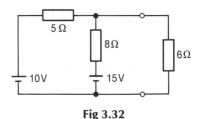

Fig 3.32

2 By means of the Superposition theorem, determine the value and direction of the current flow through the 10 Ω resistor in Fig. 3.33.

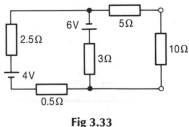

Fig 3.33

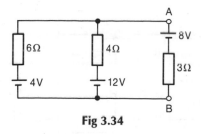

Fig 3.34

3 Use the Superposition theorem to calculate the p.d. and polarity between terminals A and B of the circuit shown in Fig. 3.34.

4 Determine the Thévénin equivalent circuit for the network to the left of terminals A and

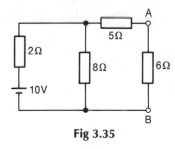

Fig 3.35

B in Fig. 3.35. Hence calculate the current flowing through the 6 Ω resistor.

5 Use Norton's theorem to calculate the current through the 6 Ω resistor of the circuit shown in Fig. 3.32.

6 Use Thévénin's theorem to solve Question 2 above.

7 For the network shown in Fig. 3.36, derive the Norton equivalent generator for the

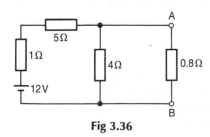

Fig 3.36

circuit to the left of terminals A and B, and hence calculate the p.d. across the 0.8 Ω resistor.

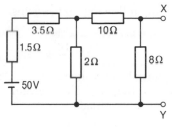

Fig 3.37

8 For the network of Fig. 3.37, determine (a) the Thévénin equivalent circuit, (b) the Norton equivalent circuit, and (c) the value of resistor, connected between terminals X and Y, which will result in the transfer of maximum power from the source.

9 Calculate the p.d. across, and current through the 3.9 Ω resistor in Fig. 3.38.

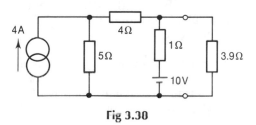

Fig 3.38

10 For the circuit of Fig. 3.39, (a) using either Thévénin's or Norton's theorem, determine

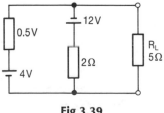

Fig 3.39

the p.d. across, and current through R_L, (b) the value to which R_L must be changed to ensure maximum power transfer into the load, and (c) the value of this maximum power.

11 An a.c. source of internal impedance 600 Ω supplies a load of impedance 150 Ω. Determine the turns ratio of the matching transformer required for maximum power transfer.

12 Express the following power gains/losses in decibels:

(a) 25, (b) 0.36, (c) 255, (d) 10^6, (e) 0.022, (f) 85

13 An amplifier has a voltage gain of 25 dB. If its output voltage is 6 V, determine the corresponding input voltage.

14 An attenuator provides a voltage attenuation of -15 dB. Calculate the resulting output voltage if a 20 V signal is applied to its input terminals.

15 Express the following powers in dBm:

(a) 100 mW, (b) 2.5 W, (c) 150 μW, (d) 0.25 mW, (e) 10 W

16 Four circuit elements are connected in series. If the power gains/losses of the individual elements are -8 dB, $+5$ dB, $+12.5$ dB, and -2.4 dB, calculate the output power if 0 dBm is applied at the input.

17 An amplifier has a voltage gain of 8 at a frequency of 100 Hz, and a voltage gain of 45 at 5 kHz. Determine the relative gain at 5 kHz to that at 100 Hz, expressing your answer in dB.

18 For the attenuator pad shown in Fig. 3.40, determine (a) the output voltage, V_o, and (b) the attenuation, in dB.

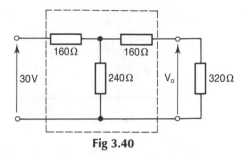

Fig 3.40

19 A 'T'-shaped attenuator is shown in Fig. 3.41. Determine the attenuation produced.

20 Calculate the attenuation provided by the network shown within the dotted lines of Fig. 3.42. If a 2 V signal is applied to terminals A and B, calculate the voltage existing between C and D.

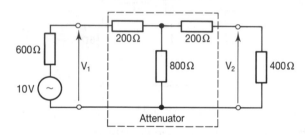

Fig 3.41

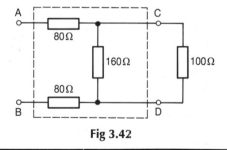

Fig 3.42

Suggested Practical Assignments

Assignment 1

To investigate the maximum power theorem.

Apparatus:
1 × a.c. signal generator (impedance 600 Ω)
1 × decade resistance box
1 × wattmeter

Method:

1 Connect the resistance box to the 600 Ω output terminals of the signal generator, with the wattmeter connected to measure the power dissipated.

2 Set the generator output to 10 V, at a frequency of 1 kHz.

3 Set the resistance to 100 Ω, and record the power dissipated.

4 Increase the resistance, in 50 Ω steps, up to 1 kΩ, noting the power dissipation at each step.

5 Plot a graph of power dissipation versus load resistance, and from your graph, determine the value of load resistance resulting in maximum load power.

6 Submit an assignment report.

Assignment 2

To investigate the response of simple low-pass and high-pass filter circuits.

Apparatus:
1 × 100 Ω resistor
1 × 0.5 μF capacitor
1 × voltmeter (with frequency response up to 6 kHz)
1 × a.c. signal generator

Method:

1 Connect the circuit of Fig. 3.43.

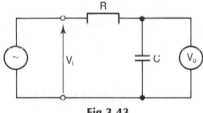

Fig 3.43

2 Set the signal generator output (V_i) to 10 V, at a frequency of 50 Hz, and measure the corresponding voltage developed across the capacitor (V_o).

3 Set the frequency to 100 Hz, ensuring that V_i is 10 V. Record the value of V_o.

4 Increase the frequency, in 100 Hz steps, up to 1 kHz, and record the value of V_o at each step. IMPORTANT: ensure that V_i is maintained at 10 V throughout the exercise.

5 Increase the frequency, in 1 kHz steps, from 1 kHz to 4 kHz, noting the value of V_o at each step.

6 On completion of the practical exercise, plot a graph of (V_o/V_i) versus frequency.

7 Transpose the positions of the resistor and capacitor in the circuit, i.e. V_o will now exist across the resistor.

8 Set the frequency to 500 Hz, with V_i still at 10 V, and record the value of V_o.

9 Increase the frequency, in 500 Hz steps, up to 4 kHz, recording the value of V_o at each step.

10 Plot a graph of (V_o/V_i) versus frequency.

11 Submit an assignment report.

Assignment 3

To observe and check the operation of an attenuator network.

Apparatus:
1 × decade step attenuator
1 × digital voltmeter
1 × variable d.c. psu

Method:

1 Connect the psu to the input terminals of the step attenuator, and the DVM to the output terminals of the attenuator.

2 Set the psu output to a convenient value. Adjust and note the attenuator setting, and record the attenuator output voltage (V_o), for a known input voltage (V_i).

3 Maintaining the original attenuation, set the input voltage to three more different values, and record the corresponding output voltages.

4 Repeat the procedures of paragraphs (2) and (3) above, for two more different attenuator settings.

5 For each set of recorded values of V_i and V_o, calculate the attenuation achieved (in decibels), and compare these results to the attenuator settings applied.

▪ 4 D.C. Transients ▪

This chapter explains the response of capacitor-resistor, and inductor-resistor circuits, when they are connected to and disconnected from, a d.c. supply.

On completion of the chapter, you should be able to:

1] Explain how the current and capacitor voltage in a series *C-R* circuit varies with time, when connected to/disconnected from a d.c. supply.

2] Explain how the current through, and p.d. across an inductor in a series *L-R* circuit varies with time, when connected to/disconnected from a d.c. supply.

3] Define the term time constant for both types of above circuits.

4] Carry out calculations in order to predict the value of current and/or voltage at any instant of time.

5] Carry out calculations in order to predict the time taken for the current and/or voltage to reach a specified value.

6] Demonstrate the affect of circuit time constant on rectangular waveforms, and relate these affects to the concepts of simple integration and differentiation.

4.1 Capacitor–Resistor Series Circuit (Charging) ——

Before dealing with the charging process for a *C-R* circuit, let us firstly consider an analogous situation. Imagine that you need to inflate a 'flat' tyre with a foot pump. Initially it is fairly easy to pump air into the tyre. However, as the air pressure inside the tyre builds up, it becomes progressively more difficult to force more air in. Also, as the internal pressure builds up, the rate at which air can be pumped in decreases. Comparing the two situations, the capacitor (which is to be charged) is analogous to the tyre; the d.c. supply behaves like the pump; the charging current compares to the air flow rate; and the p.d. developed between the plates of the capacitor has the same effect as the tyre pressure. From these comparisons we can conclude that as the capacitor voltage builds up, it reacts against the emf of the

supply, so slowing down the charging rate. Thus, the capacitor will charge at a non-uniform rate, and will continue to charge until the p.d. between its plates is equal to the supply emf. This last point would also apply to tyre inflation, when the tyre pressure reaches the maximum pressure available from the pump. At this point the air flow into the tyre would cease. Similarly, when the capacitor has been fully charged, the charging current will cease.

Let us now consider the *C-R* charging circuit in more detail. Such a circuit is shown in Fig. 4.1. Let us assume that the capacitor is initially fully discharged, i.e. the p.d. between its plates (v_C) is zero, as will be the charge, q. Note that the lowercase letters v and q are used because, during the charging sequence, they will have continuously changing values, as will the p.d. across the resistor

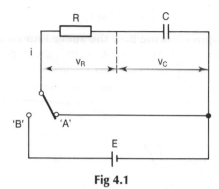

Fig 4.1

(v_R) and the charging current, i. Thus these quantities are said to have *transient* values.

At some time $t = 0$, let the switch be moved from position 'A' to position 'B'. At this instant the charging current will start to flow. Since there will be no opposition offered by capacitor p.d. ($v_C = 0$), then only the resistor, R, will offer any opposition. Consequently, the initial charging current (I_o) will have the maximum possible value for the circuit. This initial charging current is therefore given by:

$$I_o = \frac{E}{R} \text{ amp} \qquad (4.1)$$

Since we are dealing with a series d.c. circuit, then the following equation must apply *at all times*:

$$E = v_R + v_C \text{ volt} \ldots \ldots [1]$$
thus, at time $t = 0$

$$E = v_R + 0$$

i.e. the full emf of E volt is developed across the resistor at the instant the supply is connected to the circuit. Since $v_R = iR$, and at time $t = 0$, $i = I_o$, this confirms equation 4.1 above.

Let us now consider the situation when the capacitor has reached its fully-charged state. In this case, it will have a p.d. of E volt, a charge of Q coulomb, and the charging current, $i = 0$. If there is no current flow then the p.d. across the resistor, $v_R = 0$, and equ. [1] is:

$$E = 0 + v_C$$

Having confirmed the initial and final values for the transients, we now need to consider how they vary, with time, between these limits. It has already been stated that the variations will be non-linear (i.e. not a straight line graph). In fact the

variations follow an *exponential* law. This fact can be proven by the following mathematical derivation. However, please note that, for a BTEC level 3 course, you are *not* required to reproduce this derivation under examination conditions.

$$E = v_R + v_C; \quad \text{where } v_R = Ri$$

therefore, $E = Ri + v_C;$ but $i = \dfrac{dq}{dt}$

so, $E = R\dfrac{dq}{dt} + v_C$

but, $q = Cv_C;$ so $\dfrac{dq}{dt} = C\dfrac{dv_C}{dt}$

hence, $E = CR\dfrac{dv_C}{dt} + v_C$

$$CR\frac{dv_C}{dt} = E - v_C$$

so $dt = CR\left(\dfrac{1}{E - v_C}\right)dv_C$

and integrating both sides of the above equation:

$$t = CR \int \left(\frac{1}{E - v_C}\right)dv_C$$

so $t = -CR \ln(E - v_C) + k \qquad \ldots \ldots [2]$

where k is the constant of integration, which is determined as follows:

when $t = 0$, then $v_C = 0$, and substituting into [2] above:

$$0 = -CR \ln E + k; \text{ hence, } k = CR \ln E$$

Thus, equ [2] becomes:

$$t = CR \ln E - CR \ln (E - v_C)$$

$$t = CR \ln\left(\frac{E}{E - v_C}\right) = CR \ln\left(\frac{1 - v_C}{E}\right)$$

therefore, $\dfrac{t}{CR} = \ln\left(\dfrac{1 - v_C}{E}\right);$ and antilogging both sides:

$$e^{-t/CR} = 1 - \frac{v_C}{E}$$

$$\frac{v_C}{E} = 1 - e^{-t/CR}$$

hence, $v_C = E(1 - e^{-t/CR})$ volt

From the above equation it may be seen that the capacitor voltage, v_C, changes exponentially with time. Also, since we know that the capacitor voltage increases from zero to E volt, then during the transient period it follows an exponential growth.

The term CR in the exponent for e is known as the time constant, τ (greek letter tau), of the system. It may appear strange that the product of capacitance and resistance yields a result having units of time. This may be justified by considering a simple dimensional analysis, as follows.

$$C = \frac{Q}{V} = \frac{It}{V} \quad \text{and} \quad R = \frac{V}{I}$$

so, $CR = \dfrac{It}{V} \times \dfrac{V}{I} = t$ seconds

The time constant of the capacitor-resistor arrangement may be defined as the time that it would take the capacitor to become fully charged, *if* it continued to charge at the *initial* rate.

Since $q = Cv_C$, then it follows that the charge on the capacitor also grows exponentially, from zero to Q coulomb. Thus, the equations for capacitor voltage and charge are both of the form:

$$v_C = E(1 - e^{-t/\tau}) \text{ volt} \tag{4.2}$$

$$\text{and } q = Q(1 - e^{-t/\tau}) \text{ coulomb} \tag{4.3}$$

As the applied emf, E, must at all times be equal to the sum of the p.d.s across the capacitor and the resistor, then it follows that the resistor voltage must *decay* exponentially, from E volt to zero. The equation for this p.d. is:

$$v_R = E \cdot e^{-t/\tau} \text{ volt} \tag{4.4}$$

Similarly, since the p.d. across the resistor is directly proportional to the current flowing through it, then the charging current must decay exponentially from I_o to zero. The equation for the circuit current is therefore:

$$i = I_o \cdot e^{-t/\tau} \text{ amp} \tag{4.5}$$

The graphs showing how the above four quantities vary, after the supply is connected, are Figs. 4.2 to

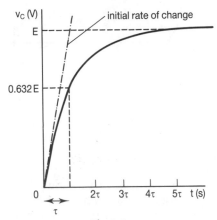

Fig 4.2

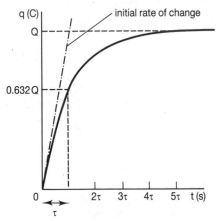

Fig 4.3

4.5 inclusive. The following points of interest arise from these graphs.

(i) For exponential growth, the quantity concerned reaches 63.2% of its final value after the elapse of one time constant.

(ii) For exponential decay, the quantity concerned falls to 36.8% of its initial value after the elapse of one time constant.

(iii) After five time constants ($t = 5\tau$ seconds), the quantity concerned has reached a value within 0.67% of its final value, e.g. $v_C = 0.993 E$ volt.

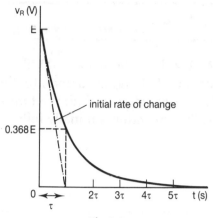

Fig 4.4

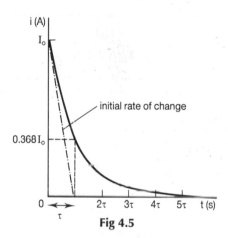

Fig 4.5

Theoretically, it is a feature of an exponential function that the capacitor voltage can never actually achieve a value of E volt. However, for practical purposes, it is assumed that the capacitor reaches its fully charged state in a time of 5τ seconds.

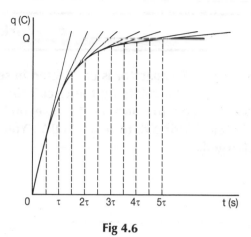

Fig 4.6

(iv) Considering *any* point on the graph, it would take one time constant for that quantity to reach its final value *if* it continued to change at the same rate as at that point. Thus the exponential graph may be considered as being formed from an infinite number of straight lines, each of which represents the slope at a particular instant of time. This is illustrated in Fig. 4.6.

Points (i) to (iii) above may be confirmed from the relevant equations, as follows:

When $t = \tau$ seconds (after the elapse of one time constant), for an exponential growth:

$$v_C = E(1 - e^{-1}) = E(1 - 0.368) = 0.632\,E$$

and for an exponential decay:

$$i = I_o e^{-1} = 0.368\,I_o$$

and when $t = 5\tau$ seconds:

$$v_C = E(1 - e^{-5}) = E(1 - 6.74 \times 10^{-3}) = 0.993\,E$$

■ Worked Example 4.1 ■

 An 8 μF capacitor is connected in series with a 0.5 MΩ resistor, across a 200 V d.c. supply. Calculate (a) the circuit time constant, (b) the initial charging current, (c) the p.d.s across the capacitor and resistor 6 seconds after the supply is connected, and (d) the time taken for the capacitor p.d. to reach 160 V. You may assume that the capacitor is initially fully discharged.

A

$C = 8 \times 10^{-6}$ F; $R = 0.5 \times 10^{6}$ Ω; $E = 200$ V

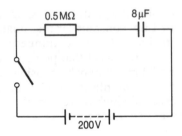

Fig 4.7

(a) $\tau = CR$ second $= 8 \times 10^{-6} \times 0.5 \times 10^{6}$

so $\tau = 4$ s **Ans**

(b)
$$I_{o} = \frac{E}{R} \text{ amp} = \frac{200}{0.5 \times 10^{6}}$$

therefore $I_{o} = 400$ μA **Ans**

(c) $v_{C} = E(1 - e^{-t/\tau})$ volt $= 200(1 - e^{-6/4})$

$v_{C} = 155.37$ V **Ans**

$v_{R} = Ee^{-t/\tau}$ volt $= 200e^{-6/4}$

so $v_{R} = 44.63$ V **Ans**

Or, more simply, $v_{R} = E - v_{C}$ volt
$$= 200 - 155.37$$

so $v_{R} = 44.63$ V as before

(d) $v_{C} = E(1 - e^{-t/\tau})$ volt

and to find the time t, first transpose the equation to make the term $e^{-t/\tau}$ the subject:

$$\frac{v_{C}}{E} = 1 - e^{-t/\tau}$$

so, $e^{-t/\tau} = 1 - \dfrac{v_{C}}{E}$

$$e^{-t/4} = 1 - \frac{160}{200} = 0.2$$

and taking logs (to the base e) of both sides:

$$-\frac{t}{4} = \ln 0.2 = -1.609$$

hence $t = 4 \times 1.609 = 6.44$ s **Ans**

Note: It may have been slightly simpler to use the fact that when $v_{C} = 160$ V, then $v_{R} = 40$ V, and then use the equation

$$v_{R} = Ee^{-t/\tau}.$$

It is left to the reader to carry out this calculation to verify the answer to part (d) above.

4.2 Capacitor-Resistor Series Circuit (Discharging) —

Consider the circuit of Fig. 4.1, where the switch has been in position 'B' for sufficient time to allow the charging process to be completed. Thus the charging current will be zero, the p.d. across the resistor will be zero, the p.d. across the capacitor will be E volt, and it will have stored a charge of Q coulomb.

At some time $t = 0$, let the switch be moved back to position 'A'. The capacitor will now be able to discharge through resistor R. The general equation for the voltages in the circuit will still apply.

In other words $E = v_R + v_C$

but, at the instant the switch is moved to position 'A', the source of emf is removed. Applying this condition to the general equation above yields:

$$0 = v_R + v_C; \text{ where } v_C = E \text{ and } v_R = I_oR$$

so $0 = I_oR + E$

hence $I_o = -\dfrac{E}{R}$ amp $\qquad\qquad$ (4.6)

This means that the initial discharge current has the same value as the initial charging current, but (as you would expect) it flows in the opposite direction.

Since the capacitor is discharging, then its voltage will decay from E volt to zero; its charge will decay from Q coulomb to zero; and the discharge current will also decay from I_o to zero. The equations for all the quantities will be as follows:

$$v_C = Ee^{-t/\tau} \text{ volt} \qquad\qquad (4.7)$$
$$q = Qe^{-t/\tau} \text{ coulomb} \qquad\qquad (4.8)$$

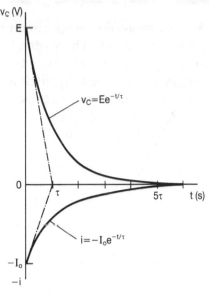

Fig 4.8

$$i = -I_oe^{-t/\tau} \text{ amp} \qquad\qquad (4.9)$$
$$v_R = Ee^{-t/\tau} \text{ volt} \qquad\qquad (4.10)$$

The graphs for v_C and i are shown in Fig. 4.8.

Note: The time constant for the C-R circuit was defined previously in terms of the capacitor charging. However, a time constant also applies to the discharge conditions also. It is therefore better to define the time constant in a more general manner, as follows:

The time constant of a circuit is the time that it would have taken for any transient variable to change, from one steady state to a new steady state, if it had maintained its original rate of change.

▪ Worked Example 4.2 ▪

Q A *C-R* charge/discharge circuit is shown in Fig. 4.9. The switch has been in position '1' for a sufficient time to allow the capacitor to become fully discharged.
(a) If the switch is now moved to position '2', calculate the capacitor p.d. after 0.3 s.
  (b) If after the 0.3 s the switch is now returned to position '1', calculate the value of discharge current and capacitor p.d. after a further period of 0.3 s.

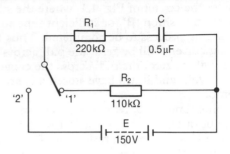

Fig 4.9

A

$C = 0.5\ \mu F;\ R_1 = 220\ k\Omega;\ R_2 = 110\ k\Omega;\ E = 150\ V$

(a) When charging, only resistor R_1 is connected in series with the capacitor, so R_2 may be ignored.

$\tau = CR_1$ seconds $= 0.5 \times 10^{-6}$
$\times 220 \times 10^3$

so $\tau = 0.11$ s

$v_C = E(1 - e^{-t/\tau})$ volt

$= 150(1 - e^{-0.3/0.11})$

hence $v_C = 140.2$ V **Ans**

(b) When discharging, both R_1 and R_2 are connected in series with the capacitor, so their combined resistance $R = R_1 + R_2$, will determine the discharge time constant.

$\tau = CR_1$ seconds $= 0.5 \times 10^{-6}$
$\times 220 \times 10^3$

so, $\tau = 0.16$ s

Also, the capacitor will be discharging from a voltage of 140.2 V. Thus, the initial discharge current will be:

$$I_o = \frac{140.2}{330 \times 10^3}\ amp$$

$$= 424.85\ \mu A$$

Therefore after 0.3 s of the discharge cycle, the current will be:

$i = I_o.e^{-t/\tau}$ amp $= 424.85\ e^{-0.3/0.16}\ \mu A$

hence, $i = 65.15\ \mu A$ **Ans**

$v_C = 140.2e^{-0.3/0.16}$ volt

so $v_C = 21.5$ V **Ans**

Note: It should be obvious from equations (4.7) and (4.10) that, in the discharge circuit, the p.d. across the capacitor and the circuit resistance must be the same value. Thus the last part of (b) could

have been obtained thus:

$$v_C = v_R = iR = 65.15 \times 10^{-6} \times 330 \times 10^3$$

hence, $v_C = 21.5$ V as above.

4.3 Inductor-Resistor Series Circuit (Connection to Supply)

Consider the circuit of Fig. 4.10. At some time $t = 0$, the switch is moved from position 'A' to position 'B'. The connection to the supply is now

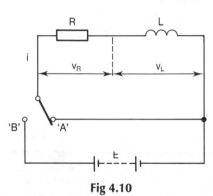

Fig 4.10

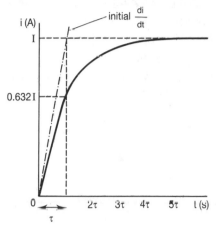

Fig 4.11

complete, and current will start to flow, increasing towards its final steady value. However, whilst the current is *changing* it will induce a back-emf across the inductor, of e volt. From electromagnetic induction theory we know that this induced emf will have a value given by:

$$e = -L\frac{di}{dt} \text{ volt}$$

Being a simple series circuit, Kirchhoff's voltage law will apply, such that the sum of the p.d.s equals the applied emf. Also, since we are considering a perfect inductor (the resistor shown may be considered as the coil's resistance), the p.d. across the inductor will be exactly equal but opposite in polarity to the induced emf.

Therefore, $v_L = -e = L\frac{di}{dt}$ volt

hence, $E = v_R + v_L$ volt

or, $E = iR + L\frac{di}{dt}$ volt .... [1]

Comparing this equation with that for the *C-R* circuit, it may be seen that they are both of the same form. Both contain a first-derivative term (a 'd/dt' term) and a 'constant' term. Using the

analogy technique, we can conclude that both systems will respond in a similar manner. In the case of the *L-R* circuit, the *current* will increase from zero to its final steady value, following an exponential law.

At the instant that the switch is moved from 'A' to 'B' ($t = 0$), the current will have an instantaneous value of zero, but it *will* have a certain rate of change, di/dt amp/s. From equ [1] above, this initial rate of change can be obtained, thus:

$$E = 0 + L\frac{di}{dt}$$

so, initial $\frac{di}{dt} = \frac{E}{L}$ amp/s $\qquad$ (4.11)

When the current reaches its final steady value, there will be no back-emf across the inductor, and hence no p.d. across it. Thus the only limiting factor on the current will then be the resistance of the circuit. The final steady current is therefore given by:

$$I = \frac{E}{R} \text{ amp} \qquad (4.12)$$

The time constant of the circuit is obtained by dividing the inductance by the resistance.

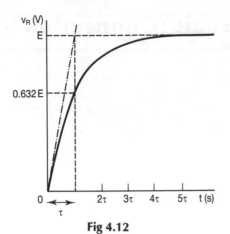

Fig 4.12

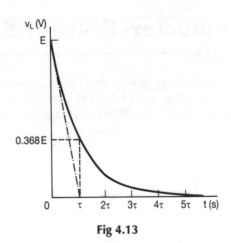

Fig 4.13

Thus $$\tau = \frac{L}{R} \text{ seconds} \qquad (4.13)$$

The above equation may be confirmed by using a simple form of dimensional analysis, as follows.

$$\text{In general, } V = \frac{LI}{t}; \text{ so } L = \frac{Vt}{I}$$

$$\text{and } R = \frac{V}{I}$$

$$\text{therefore, } \frac{L}{R} = \frac{Vt}{I} \times \frac{I}{V} = t \text{ seconds}$$

The time constant of the circuit may be defined in the general terms given in the *'Note'*, in the previous section, dealing with the *C-R* circuit.

The rate of change of current will be at its maximum value at time $t = 0$, so the p.d. across the inductor will be at its maximum value at this time. This p.d. therefore decays exponentially from E volt to zero. The graphs for i, v_R, and v_L are shown in Figs. 4.11 to 4.13 respectively.

▪ Worked Example 4.3 ▪

 The field winding of a 110 V, d.c. motor has an inductance of 1.5 H, and a resistance of 220 Ω. From the instant that the machine is connected to a 110 V supply, calculate (a) the initial rate of change of current, (b) the final steady current, (c) the current 10 ms after connection, and (d) the time taken for the current to reach its final steady value.

A

$E = 110$ V; $L = 1.5$ H; $R = 220$ Ω; $t = 10$ ms

The circuit diagram is shown in Fig. 4.14.

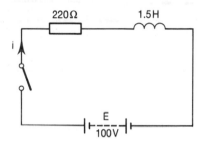

220Ω   1.5H

i

E
100V

Fig 4.14

(a)

$$\text{initial } \frac{di}{dt} = \frac{E}{L} \text{ amp/s} = \frac{110}{1.5}$$

$$\text{so, initial } \frac{di}{dt} = 73.33 \text{ A/s } \textbf{Ans}$$

(b)

$$\text{final current, } I = \frac{E}{R} \text{ amp} = \frac{110}{220}$$

$$\text{therefore, } I = 0.5 \text{ A } \textbf{Ans}$$

(c)

$$\tau = \frac{L}{R} \text{ second} = \frac{1.5}{220}$$

hence, $\tau = 6.82$ ms

$$i = I(1 - e^{-t/\tau}) \text{ amp}$$

$$= 0.5(1 - e^{-10/6.82})$$

therefore, $i = 0.385$ A **Ans**

(d) Since the system takes approximately 5τ seconds to reach its new steady state, then the current will reach its final steady value in a time:

$$t = 5 \times 6.82 \text{ ms} = 34.1 \text{ ms } \textbf{Ans}$$

4.4 Inductor-Resistor Series Circuit (Disconnection) –

Fig. 4.15 shows such a circuit, connected to a d.c. supply. Assume that the current has reached its final steady value of I amps. Let the switch now be returned to position 'A' (at time $t = 0$). The current will now decay to zero in an exponential manner. However, the decaying current will induce a back-emf across the coil. This emf must oppose the change of current. Therefore, the decaying current will flow *in the same direction* as the original steady current. In other words, the back-emf will try to maintain the original current

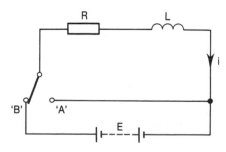

R   L

i

'B'   'A'

E

Fig 4.15

flow. The graph of the decaying current, with respect to time, will therefore be as shown in Fig. 4.16. The time constant of the circuit will, of course, still be L/R second, and the current will decay from a value of $I = E/R$ amp. The initial rate of decay will also be E/L amp/s.

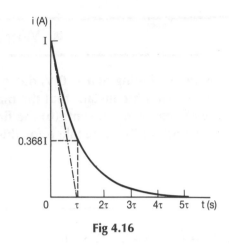

Fig 4.16

***

■ **Worked Example 4.4** ■

***

Q Fig. 4.17 represents a relay coil that is energised/de-energised by means of a 2-pole switch, which has been in the position shown for some considerable time. In order for the relay to close the contact, the coil current must be 0.8 A. When de-energised, the contact opens when the coil current has fallen to 0.5 A. Calculate (a) the time taken for the contact to close, and (b) the time taken for the contact to open, if the switch is left in position '2' for 25 ms, before it is returned to position '1'. Sketch a graph showing the variation of current, versus time, for a total of 35 ms after the supply is first connected, showing all principal values.

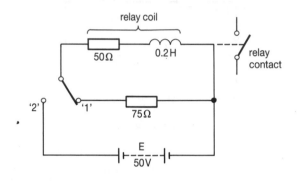

Fig 4.17

$L = 0.2$ H; $R_1 = 50\ \Omega$; $R_2 = 125\ \Omega$; $E = 50$ V

$i_1 = 0.8$ A; $i_2 = 0.5$ A

(a) For current growth, $\tau = \dfrac{L}{R_1}$ second $= \dfrac{0.2}{50}$

so, $\tau = 4$ ms

Final steady current, $I = \dfrac{E}{R_1}$ amp $= \dfrac{50}{50} = 1$ A

Now, $i_1 = I(1 - e^{-t/\tau})$ amp

$\dfrac{i_1}{I} = 1 - e^{-t/\tau}$

$$e^{-t/\tau} = 1 - \frac{i_1}{I}$$

$$\frac{-t}{\tau} = \ln\left(1 - \frac{i_1}{I}\right)$$

therefore, $-t = \tau \ln\left(1 - \frac{i_1}{I}\right)$ second

$$= 4 \ln (1 - 0.8)$$

hence, $t = 6.438$ ms **Ans**

(b) The current will grow to its final value in $5\tau = 20$ ms. Since the switch is left in position '2' for 25 ms, then the current will be at its steady value of 1 A when the switch is returned to position '1'. Hence, the initial decay current will be $I - 1$ A. However, in position '1', the decay current will flow through both the 50 Ω and the

75 Ω resistor. That is to say, the total resistance, $R_2 = 125$ Ω. This means that the decay current time constant is given by:

$$\tau = \frac{L}{R_2} \text{ second} = \frac{0.2}{125} = 1.6 \text{ ms}$$

Also, $i_2 = Ie^{-t/\tau}$ amp

$$\frac{i_2}{I} = e^{-t/\tau}$$

$$-t = \tau \ln \frac{i_2}{I} = 1.6 \ln 0.5 \text{ ms}$$

therefore, $t = 1.11$ ms **Ans**

The current decays to zero in approximately 8 ms, and the graph of current versus time will be as shown in Fig. 4.18.

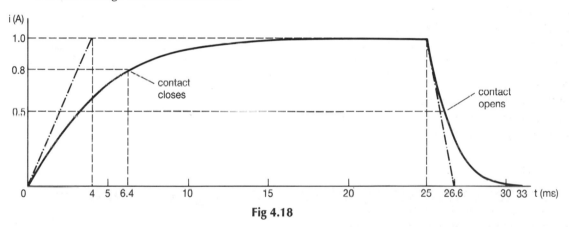

Fig 4.18

4.5 Affect of Time Constant on Alternating Waveforms

Consider a series *C-R* circuit, with a rectangular waveform applied to the input terminals, as in Fig. 4.19. The p.d.s developed across the resistor and capacitor are monitored on a double-beam oscilloscope. If the resistor is adjusted to a low value, then the circuit time constant ($\tau = CR$) will be short. Let us assume that, in this case, τ is very much less than the half-period ($T/2$) of the input waveform. Since the time constant is very short,

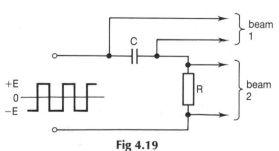

Fig 4.19

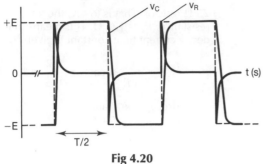

Fig 4.20

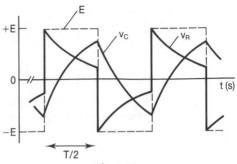

Fig 4.21

then the capacitor can charge and discharge rapidly. The waveform for the p.d. developed across C is shown in Fig. 4.20. Also shown is the p.d. developed across R. Notice that if these two waveforms are added, the input rectangular waveform is reproduced. This confirms the fact that $E = v_R + v_C$ volt.

Let the time constant now be increased, by increasing the value of R. Let us also assume that the new time constant is sufficiently long so as to prevent the capacitor reaching its fully charged state of E volt. The affect on the p.d.s is illustrated in Fig. 4.21.

Finally, let the time constant be further increased, to the point where τ is very much greater than $T/2$. The resulting waveforms are

shown in Fig. 4.22. In this situation the capacitor charges and discharges so slowly that the waveform of v_C shows only the initial rates of change of v_C, which are a close approximation to straight lines.

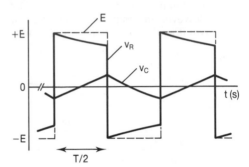

Fig 4.22

4.6 Differentiating and Integrating Networks

Consider the network of Fig. 4.23. The input waveform is a rectangular wave, and the network

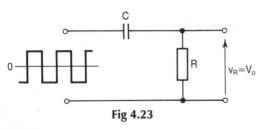

Fig 4.23

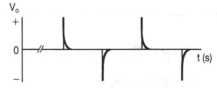

Fig 4.24

output is taken across the resistor. Let the circuit time constant be very short compared with the half-period of the input. To meet this criterion, the 'rule of thumb' is that τ is less than, or equal to one tenth of the input half-period, i.e. $\tau \leqslant T/20$.

The resulting output waveform is shown in Fig. 4.24.

Now, the mathematical process of differentiation is merely a means of determining the rate of change of a quantity. For example, velocity is the rate of change of distance with time. Expressed mathematically this is written as $v = \mathrm{d}s/\mathrm{d}t$.

A *perfect* rectangular waveform has vertical rising and falling edges, and horizontal 'tops'. However, vertical edges implies instantaneous changes from one value to another, i.e. a change in zero time. Such a change is a physical impossibility, since the rate of change would be *infinite*. In practice, the rising and falling edges of a rectangular waveform are very steep, but cannot be absolutely vertical. The 'tops' of the waveform can be horizontal, since this is a zero rate of change. However, if a perfect rectangular waveform was differentiated, the result would be a series of vertical 'spikes' of infinite length and of zero width (occupying zero time). This is illustrated in Fig. 4.25. If we now

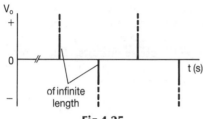

Fig 4.25

compare this waveform with that of Fig. 4.24, it may be seen that the output voltage of the network is a very close approximation to the differential of the input waveform. Indeed, the shorter the circuit time constant, the 'better' the differentiating effect.

This simple form of differentiating network is used in some electronic circuits to convert a relatively wide rectangular pulse into a 'spike'. This technique is often used to generate the trigger pulses to electronic timing circuits, such as a device known as a 555 timer.

Let us now transpose the places of C and R in the circuit, and also alter their values to provide a very long time constant. In this case, the 'rule of thumb' is that τ is equal to, or greater than, ten times the half-period of the input waveform, i.e. $\tau \geqslant 5T$ seconds. The circuit arrangement is shown in Fig. 4.26, and the resulting output waveform is

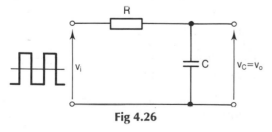

Fig 4.26

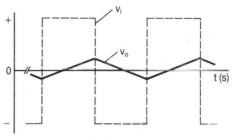

Fig 4.27

shown in Fig. 4.27. If a rectangular waveform is integrated, then a triangular waveform, of reduced amplitude, results. Thus this circuit is an integrating network. This reduction of amplitude may be confirmed by considering the integration of a sinewave, as follows.

$$\text{Let the input, } v_i = V_{im} \sin \omega t \text{ volt}$$

$$\text{and the output, } v_o = \int v_i \, dt$$

$$\text{then, } v_o = \int V_{im} \sin \omega t \, dt \text{ volt}$$

$$\text{hence, } v_o = -\frac{V_{im}}{\omega} \cos \omega t \text{ volt}$$

i.e. the amplitude is *reduced* by a factor of ω.

The integrating action of the circuit may be more easily understood by considering a sinusoidal input, as in Fig. 4.28. In order to achieve a long

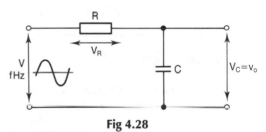

Fig 4.28

time constant we could have large values for both the resistor and the capacitor. Since X_C is proportional to $1/C$, then for a large value capacitor, X_C will be very much less than R. The resulting phasor diagram is shown in Fig. 4.29. From this diagram it may be seen that $V \approx V_R$, and

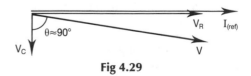

Fig 4.29

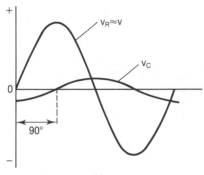

Fig 4.30

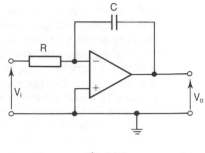

Fig 4.31

the 'output', V_C, lags V by a large angle θ, which is approaching 90°. The corresponding waveform diagram is shown in Fig. 4.30. From this it may be seen that the output, V_C, has a very small amplitude, and is a negative cosine wave. This is precisely the mathematical result shown above, which confirms that the circuit acts as an integrator.

It is left to the reader to confirm the differentiating action for a sinusoidal input, using the appropriate phasor and waveform diagram. Remember though, for 'good' differentiation, a very short time constant is required.

In most practical applications using an integrator the severe reduction in amplitude of the output waveform is undesirable. To overcome this problem an active integrator is used, which

involves the use of an operational amplifier. A circuit of such an arrangement is shown in Fig. 4.31. This form of active integrator is utilised in applications such as analogue computers (simulators), some types of digital voltmeter, and oscilloscope timebase circuits. The latter two applications require the production of a linearly changing, or ramp, voltage.

It should be noted that series *L-R* circuits will also exhibit the differentiating and integrating properties described above. However, due to the relative weight, bulk and cost of large value inductors, integrating and differentiating networks using inductors are not deliberately used to achieve these effects. Any circuit containing inductors can of course cause the attenuation and distortion to rectangular waveforms described above, but this is usually an unwanted effect.

Assignment Questions

1 A 47 µF capacitor is connected in series with a 39 kΩ resistor, across a 24 V d.c. supply. Calculate (a) the circuit time constant, (b) the values for initial and final charging current, and (c) the time taken for the capacitor to become fully charged.

2 A 150 mH inductor of resistance 50 Ω is connected to a 50 V d.c. supply. Determine (a) the initial rate of change of current, (b) the final steady current, and (c) the time taken for the current to change from zero to its final steady value.

3 An inductor of negligible resistance and inductance 0.25 H, is connected in series

with a 1.5 kΩ resistor, across a 24 V d.c. supply. Calculate (a) the current flowing after one time constant, (b) the p.d. across the inductor after two time constants, and (c) the p.d. across the resistor after three time constants.

4 An uncharged capacitor of 200 nF is connected to a 150 V d.c. supply via a 75 kΩ resistor. Determine the capacitor p.d. and circuit current 10 ms after connection.

5 An 8 µF capacitor is charged to 100 V, and then discharged through a 1.2 MΩ resistor. Calculate (a) the capacitor p.d. 2.5 s after discharging has commenced, and (b) the time taken for this voltage to fall to 20 V.

6 A 5 H inductor has a resistance R ohm. This inductor is connected in series with a 10 Ω resistor, across a 140 V d.c. supply. If the resulting circuit time constant is 0.4 s, determine (a) the value of the coil resistance, and (b) the current flowing 650 ms after connection to the supply.

7 Define the *time constant* of a capacitor–resistor series circuit.
  Such a circuit comprises a 50 μF capacitor and a resistor, connected to a 100 V d.c. supply via a switch. If the circuit time constant is to be 5 s, determine (a) the resistor value, (b) the initial charging current, (c) the time taken for the capacitor p.d. to reach 85 V, and (d) the value of the charging current 12 s after closing the switch.

8 The dielectric of a 20 μF capacitor has a resistance of 65 MΩ. This capacitor is fully charged from a 120 V d.c. supply. Calculate the time taken, after disconnection from the supply, for (a) the capacitor stored p.d. to fall to 99.8 V, and (b) the capacitor to become fully discharged.

9 A circuit consists of a 270 Ω resistor in parallel with a 4 H inductor, of resistance 30 Ω. If this combination is connected across a 100 V d.c. supply for 10 ms, and then disconnected, calculate the current flowing 5 ms after disconnection.

10 An electronic device is required to operate 1.5 s after a switch is closed. The device will operate when its input voltage (V_i) is equal to or greater than 5 V. This time delay is to be achieved by charging a capacitor via a resistor, as shown in Fig. 4.32. Determine the value to which the

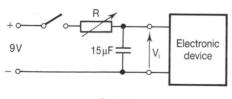

Fig 4.32

resistor must be adjusted in order to achieve the required time delay.

11 A relay coil has an inductance of 120 mH and resistance 20 Ω. When it is connected to a 50 V d.c. supply, determine (a) the initial rate of change of current, (b) the time taken for the

current to reach 1.5 A, and (c) the value of the current 4 ms after connection.

12 For the circuit shown in Fig. 4.33, determine the p.d. between terminals X and Y,

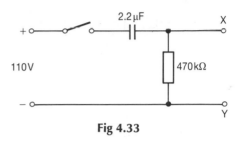

Fig 4.33

(a) before the switch is closed, (b) at the instant that the switch is closed, and (c) 0.85 s after the switch is closed.

13 The field winding of a 110 V d.c. motor has a resistance of 12 Ω, and a time constant of 2.5 s. Calculate (a) the inductance of the winding, (b) the time for the current to reach 1.8 A, and (c) the current flowing 1 s after connection to the 110 V supply.

14 For the circuit arrangement of Fig. 4.34, the capacitor is initially uncharged. At some time

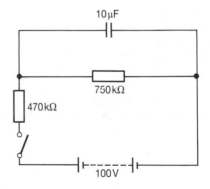

Fig 4.34

$t = 0$, the switch is closed for a period of 30 s; after which it is re-opened. (a) Sketch (to scale) the variation of capacitor current, from $t = 0$ until the capacitor is again fully discharged. Show all principal values on your graph. (b) Calculate the charge on the capacitor 10 s after connection, and

(c) the p.d. across the 750 Ω resistor 20 s after disconnection.

15 A squarewave of frequency 1 kHz and amplitude 10 V is applied to terminals A and B of the network shown in Fig. 4.35. (a) On the

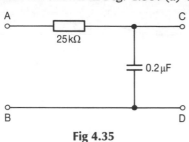

Fig 4.35

same axes, sketch the graphs of V_{AB} and V_{CD}, (b) state the mathematical function carried out by this network, and (c) indicate the effect on V_{CD} if the resistor value is reduced to 2 kΩ.

16 A squarewave of frequency 1 kHz and amplitude 5 V is applied to terminals A and B of the network shown in Fig. 4.36. (a)

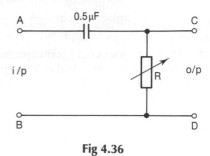

Fig 4.36

Determine the maximum allowable value for the resistor, in order for the circuit to act as a 'good' differentiator, and (b) sketch, on the same axes, the input and output waveforms under this condition.

Suggested Practical Assignments

Assignment 1

To investigate the variation of capacitor voltage and current during charge and discharge cycles.

Apparatus:
1×10 μF capacitor
1×10 MΩ resistor
1×2-pole switch
$1 \times$ d.c. power supply
$1 \times$ ammeter (microammeter)
$1 \times$ DVM (with highest possible input resistance)
$1 \times$ stopwatch

Method:

1 Connect the circuit of Fig. 4.37, and adjust the psu output to 250 V.

2 Simultaneously move the switch to position '1' and start the stopwatch.

3 Record the circuit current and capacitor p.d. at 10 s intervals, for the first 60 s.

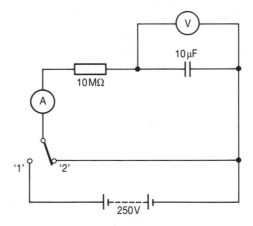

Fig 4.37

4 Continue recording the current and voltage readings, at 20 s intervals, for a further 4 minutes. Reset the stopwatch to zero. Reverse the connections to the ammeter.

5 Move the switch back to position '2', and repeat the procedures of paragraphs (3) and (4) above.

6 Plot graphs of current and capacitor p.d., versus time, for both the charging and discharging cycles.

7 Submit a complete assignment report, which should include the following:

(i) The comparison of the actual time constant (determined from the plotted graphs) to the theoretical value. Explain any discrepancy found.

(ii) Calculate several transient voltage values, and compare these with actual values plotted. Hence, justify whether or not the graphs follow an exponential change.

(iii) Explain why both the charging and discharging currents tend to 'level off' at some small value, rather than continuing to decrease to zero.

Assignment 2

To observe the affect of circuit time constant on alternating waveforms.

Apparatus:
1 × sine/squarewave signal generator
1 × variable capacitor
1 × variable resistor
1 × double-beam oscilloscope

Method:

1 Connect the circuit of Fig. 4.38, with $R = 100$ kΩ and $C = 0.1$ nF. Set the signal generator to provide a squarewave output of amplitude 5 V, at a frequency of 1 kHz.

2 Observe the two waveforms displayed on the oscilloscope, and make a note of the amplitudes, time intervals and shapes of both.

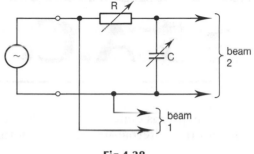

Fig 4.38

3 Select $C = 1$ nF, and repeat the above procedure.

4 Select $C = 0.1$ µF, and repeat the measurements.

5 Switch the signal generator to a sinewave output, and repeat the procedures of paragraphs (2) to (4) above.

6 Transfer the connections for beam 2 of the oscilloscope so as to monitor the voltage across the resistor.

7 Set $C = 0.01$ µF, $R = 5$ kΩ, and the signal generator to squarewave output.

8 Repeat the procedures of paragraphs (2) to (5) above.

9 Submit an assignment report, which should include scaled sketches of all the waveforms measured. Comment on the affects of different time constants on the waveforms. Identify the mathematical functions that may be performed in each case.

▪ 5 A.C. Machines ▪

This chapter is concerned with the principles of operation of transformers, a.c. generators (alternators) and a.c. motors. On completion you should be able to:

1] Apply the emf equation for a transformer.

2] Understand the losses associated with a practical transformer, and hence explain the operation of the practical device.

3] Take measurements and calculate the efficiency of a transformer under varying load conditions.

4] Explain the principle of operation of single-phase and three-phase alternators, and carry out simple calculations concerning generated emf and efficiency.

5] Explain how a polyphase supply, when connected to a polyphase stator winding, produces a rotating magnetic field.

6] Explain the principles of polyphase and single-phase synchronous and induction motors, and the losses associated with these machines.

7] Explain the various methods for starting both polyphase and single-phase motors.

8] Explain the principles of simple stepping motors.

5.1 Transformer emf Equation

Figure 5.1 shows one complete cycle of the flux waveform in the core of a transformer. From this diagram the following statements can be made:

Total change of flux
$$= \Phi_m - (-\Phi_m) = 2\Phi_m \text{ weber}$$

time taken for this change
$$= T/2 \text{ second}$$

therefore average rate of change of flux
$$= \frac{2\Phi_m}{T/2} = \frac{4\Phi_m}{T} \text{ Wb/s}$$

in general, $e = -N\mathrm{d}\phi/\mathrm{d}t$ volt, so for either winding:

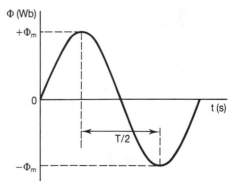

Fig 5.1

average emf induced

$$= \frac{4N\Phi_m}{T} \text{ volt, and since } T = 1/f, \text{ then}$$

average emf induced $= 4\Phi_m Nf$ volt

and since we are dealing with sinewaves, having a form factor of 1.11, the r.m.s. emf induced is given by:

$$E = 4.44\Phi_m Nf \text{ volt} \qquad (5.1)$$

Notes:

1 The above equation applies to both primary and secondary induced emfs. For the primary, read E_p and N_p, and for the secondary, E_s and N_s.

2 The minus sign from the general equation has been omitted, but of course Lenz's law still applies. Thus E_p is in antiphase to the applied voltage V_p. Since E_p and E_s are induced by the same flux, then these two emfs are in phase with each other. Finally, since the secondary terminal voltage, V_s is produced by E_s, then V_s is antiphase to V_p. This confirms that the output of a transformer is phase inverted compared with its input.

■ Worked Example 5.1 ■

Q The primary winding of a single-phase transformer is connected to a 240 V, 50 Hz supply. The secondary winding has 1500 turns. If the peak value of the core flux is 1.965 mWb, determine (a) the number of turns on the primary winding, (b) the secondary induced emf, and (c) the core cross-sectional area if the flux density has a maximum value of 0.47 T.

A

$V_p = E_p = 240$ V; $f = 50$ Hz; $N_s = 1500$;
$\Phi_m = 0.001965$ Wb; $B_m = 0.47$ T

(a) $\qquad E_p = 4.44\Phi_m N_p f$ volt

therefore $N_p = \dfrac{E_p}{4.44\Phi_m f}$

$\qquad = \dfrac{240}{4.44 \times 0.001965 \times 50}$

hence $N_p = 550$ **Ans**

(b) $\qquad E_s = 4.44\Phi_m N_s f$ volt
$\qquad\quad = 4.44 \times 0.001965 \times 1500 \times 50$

hence $E_s = 654.3$ V **Ans**

(c) $\qquad B_m = \dfrac{\Phi_m}{A}$ tesla, so $A = \dfrac{\Phi_m}{B_m}$ metre^2

therefore $A = \dfrac{0.001965}{0.47} = 41.81$ cm^2 **Ans**

5.2 The Practical Transformer

The basic principles of operation for an ideal transformer were dealt with in Volume 1. These principles were briefly reiterated in Chapter 3 of this book, under the heading of maximum power transfer, where the principle of a matching transformer was described.

In the case of the matching transformer, it was stated that when a load is connected to the secondary winding, then a primary current is drawn. This must occur, since if the secondary supplies power to the load, then this power must be drawn from the a.c. supply connected to the primary. With the ideal machine, this input power will be the same value as that supplied to the secondary load. The phasor diagram for an ideal transformer, supplying a purely resistive load, is shown in Fig. 5.2. Note that, as with any phasor

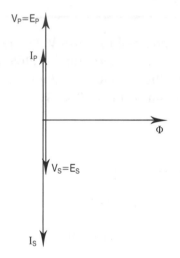

Fig 5.2

diagram, a reference phasor must be chosen. In the case of a transformer, the only quantity that is common to both the primary and secondary is the magnetic flux in the core. For this reason, the flux, Φ, is the reference phasor. Also note that a $2:1$ step-down transformer has been chosen; hence V_p is twice V_s, and I_s is twice I_p. In addition, $E_p = V_p$ and $E_s = V_s$.

If the load was now disconnected from the secondary, then I_s would be zero, and so too would I_p. However, the primary winding is still connected to the a.c. supply, so for a *practical* machine, some

primary current must still be drawn. This is known as the primary no-load current, I_o. Thus, on no-load, the transformer is drawing power from the supply, even though it is not supplying any current to a load.

This no-load primary current is flowing through a highly inductive circuit (the primary winding), the resistance of which is designed to be as small as possible. Consequently, I_o will lag V_p by a large angle, ϕ_o. This current has two components, shown as I_{mag} and I_c in Fig. 5.3. I_{mag} is the

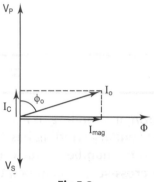

Fig 5.3

magnetising component of the current, which is responsible for producing the flux in the core. The other component, I_c, is in phase with V_p, and the product of these two quantities accounts for power loss in the transformer under these conditions. Since the resistance of the windings is small, and the no-load primary current is also relatively small, then the $I_o^2 R_p$ loss is insignificant. So what then is responsible for the power consumption? The answer to this question lies in the fact that the core flux is, of course, an alternating quantity. Consequently, the transformer core is being taken through continuous magnetisation cycles. This will result in both hysteresis and eddy current losses in the core. The affect of these losses is to cause unwanted heating of the 'iron' core, and are usually referred to collectively as the *iron losses*. The iron losses are kept to a minimum by making the core from thin laminations of a 'soft' magnetic material. The iron losses may be calculated from the equation shown below.

$$\text{Iron loss, } P_{Fe} = V_p I_o \cos \phi_o \text{ watt} \qquad (5.2)$$

Let us now consider the practical transformer connected to a partly resistive, partly inductive load, as illustrated in Fig. 5.4. This will result in a

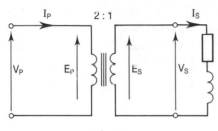

Fig 5.4

secondary current, I_s, and a corresponding primary 'balancing' current, I_p'. However, there will still be the current I_o flowing in the primary, since the core flux and core losses still exist. Hence the total primary current I_p must consist of the combination of these two currents. The appropriate phasor diagram is shown in Fig. 5.5. In this diagram, the

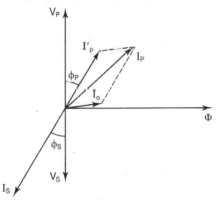

Fig 5.5

relative size of I_o has been exaggerated simply for clarity. The practical consequence is that the primary phase angle, ϕ_p, is very nearly the same as that for the secondary, ϕ_s, and the current ratio is still taken to be the inverse of the turns ratio.

We have now seen that when the transformer is connected to a secondary load, the primary

balancing current flows, so the total primary current increases in sympathy with the load current. The resistances of the two windings will now cause a more significant power loss, known as the *copper losses*. In addition, the impedance of each winding will introduce internal voltage drops, such that $E_p < V_p$ and $V_s < E_s$. This effect is illustrated in the phasor diagram of Fig. 5.6.

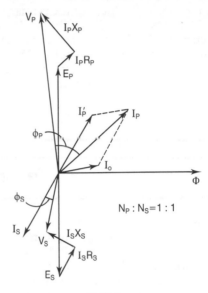

Fig 5.6

Note: This phasor diagram is included here merely for completeness. You are not expected to be able to reproduce this diagram under examination conditions.

The copper loss may be calculated from:

$$\text{Copper loss, } P_{Cu} = I_p^2 R_p + I_s^2 R_s \text{ watt} \quad (5.3)$$

It should be borne in mind that despite the losses described, the transformer is the most efficient of all the electrical machines. Efficiency figures in the order of 95% to 99% are typical. The main reason for this high efficiency is the absence of any moving parts.

5.3 Measurement of Transformer Losses

The iron and copper losses may be determined by conducting two simple tests, known as the open-circuit and short-circuit tests. Once these losses are established, the efficiency can be calculated.

■ Open-Circuit Test ■

This test is used to measure the iron losses of the transformer. As the title implies, the secondary is left on open-circuit. The primary is connected to its normally rated supply voltage, as shown in Fig. 5.7. Under this no-load condition, the core flux will attain its normal value, but the primary current (I_o) will be about 5% of the full-load primary current (I_p). Under these circumstances the primary copper loss will be only 0.25% of the full-load value. Thus the wattmeter reading may be taken as the measurement of the core iron losses.

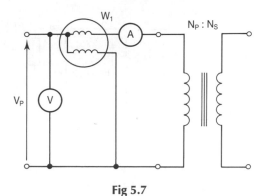

Fig 5.7

■ Short-Circuit Test ■

This test is used to measure the copper losses when the transformer is supplying its rated full-load secondary current. Although this test is simple to apply, considerable care needs to be exercised in order to protect both the transformer and the supply from damage. The circuit arrangement is shown in Fig. 5.8, and the procedure is as follows.

> A **variac** is simply a transformer having only one winding. The output is tapped off from a sliding contact. This device therefore acts as an a.c. potentiometer.

By means of the **variac** the primary voltage is very carefully increased, from zero, until the ammeter connected between the secondary terminals indicates the normally rated full-load output current. Under these conditions the primary voltage will be about 5% of its normal value, as will be the core flux. Since the iron loss is approximately proportional to the square of the flux, then this loss will be about 0.25% of its normal value. Hence the wattmeter reading may

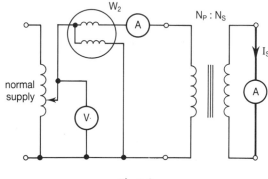

Fig 5.8

be taken as the transformer copper losses.

If the ammeter and voltmeter readings are noted when the above two tests are carried out, then the no-load and full-load power factors and phase angles may be determined.

From the open-circuit tests: $\cos \phi_o = \dfrac{P_1}{V_p I_o}$

From the short-circuit test: $\cos \phi = \dfrac{P_2}{V_p I_p}$

5.4 Transformer Efficiency

The efficiency of any device may be found by dividing its output power by the power supplied to it. The difference between the output and input must be the total losses of the machine. Thus, the efficiency may be expressed as:

$$\text{efficiency, } \eta = \frac{\text{output}}{\text{output} + \text{losses}}$$

The losses of the transformer are the iron and copper losses as measured by the open and short-circuit tests, and the full-load output is $V_s I_s \cos \phi$ watt. If we refer to these two forms of loss as P_{Fe} and P_{Cu} respectively, then the full-load efficiency is given by the expression:

$$\eta = \frac{V_s I_s \cos \phi}{V_s I_s \cos \phi + P_{Fe} + P_{Cu}} \times 100\% \qquad (5.4)$$

The above equation will give the efficiency under full-load conditions provided that $V_s I_s \cos \phi$ is the rated full-load output. When the transformer is supplying a smaller load its efficiency figure will not be the same value. The reason is that the copper losses, due to the resistance of the windings, will vary as the current varies. Thus the copper loss is a variable loss. On the other hand, the core losses are constant (fixed) regardless of the load supplied. The transformer will operate at its maximum efficiency when the variable loss equals the fixed loss. This is similar to the concept of maximum power transfer from a source to a load, i.e. when the load resistance (variable) equals the internal resistance (fixed) of the source. Hence, the equation for maximum efficiency is:

$$\eta_{max} = \frac{\text{output}}{\text{output} + 2 \times P_{Fe}} \times 100\% \qquad (5.5)$$

▪ Worked Example 5.2 ▪

 A single-phase transformer is rated at 10 kVA, 240 V/100 V. When tested the following results were obtained:

On open-circuit:

$$V_p = 240 \text{ V}; I_o = 2.6 \text{ A}; P_{Fe} = 200 \text{ W}$$

On short-circuit:

$$V_p = 18 \text{ V}; I_p = 100 \text{ A}; P_{Cu} = 250 \text{ W}$$

Using these results calculate (a) the no-load power factor, (b) the full-load efficiency, (c) the load at which maximum efficiency occurs, and (d) the value of maximum efficiency. Assume a load power factor of 0.8.

(a)
$$P_{Fe} = V_p I_o \cos \phi_o \text{ watt}$$

so, $\cos \phi_o = \dfrac{P_{Fe}}{V_p I_o} = \dfrac{200}{240 \times 2.6}$

hence, $\cos \phi_o = 0.321$ **Ans**

(b) Since the transformer is rated at 10 kVA and the load power factor is 0.8, then the full-load power output is $10 \times 0.8 = 8$ kW.

full-load efficiency, $\eta = \dfrac{\text{output}}{\text{output} + \text{losses}} \times 100\%$

$$= \dfrac{8}{8 + 0.2 + 0.25} \times 100\%$$

therefore, $\eta = 94.67\%$ **Ans**

(c) The copper loss is directly proportional to the square of the current, and for maximum efficiency $P_{Cu} = P_{Fe} = 200$ W

Full-load current $= 100$ A;

so $100^2 \propto 250 \text{ W} \ldots \ldots$ [1]

and for η_{max}; $I^2 \propto 200 \text{ W} \ldots \ldots$ [2]

and dividing [2] by [1] we have:

$$I = \sqrt{\dfrac{200}{250} \times 100^2} = 89.44 \text{ A}$$

$$P_o = 100 \times 89.44 \times 0.8$$
$$= 7155.4 \text{ W } \textbf{Ans}$$

(d)
$$\eta_{max} = \dfrac{\text{output}}{\text{output} + 2 \times P_{Fe}} \times 100\%$$

$$= \dfrac{7155.4 \times 100}{7155.4 + 400}$$

hence, $\eta_{max} = 94.71\%$ **Ans**

Notice that there is very little difference between the full-load and maximum efficiency figures.

5.5 Alternators

The basic principles of operation of both single-phase and three-phase alternators has already been dealt with in Chapter 2, as an introduction to three-phase systems. It is suggested that you re-read the introductory section of that chapter to familiarise yourself with the construction of these machines.

5.6 Rotor Construction

The rotor of an alternator may be of one of two types: the salient pole type or the cylindrical type. The choice of construction depends on the speed of the prime mover which is used to drive the rotor.

■ Salient pole rotor ■

This form of construction is used when the driving speed is relatively low. Since the frequency of the emf generated, $f = np$ hertz, then for a given frequency, the lower the speed the greater the number of pole pairs required. For example, if the speed of rotation is 750 rev/min, then an 8-pole rotor would be required to generate at 50 Hz. This calculation is shown below.

$$f - 50 \text{ Hz}; \; n = \frac{750}{60} \text{ rev/s}$$

$$p = \frac{f}{n} = \frac{50 \times 60}{750} = 4$$

and number of poles $= 2p = 8$.

In order to accommodate a large number of poles, the alternator needs to have a relatively large diameter. Also, since the power output of a machine is roughly proportional to its volume, this

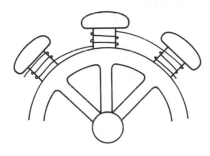

Fig 5.9

type of alternator would have a comparatively small axial length. Fig. 5.9 illustrates a portion of a salient pole rotor. The d.c. field winding would be contrawound on each alternate pole piece so as to provide alternate polarities (North–South–North etc.) at the tip of each. Each pole tip would also be well rounded to ensure a sinusoidal waveform.

■ Cylindrical rotor ■

The majority of alternators are driven by steam turbines, which are essentially high speed machines. In this situation a large number of poles is not necessary, and most large alternators of this type are two-pole machines, driven at 3000 rev/minute, resulting in a frequency of 50 Hz. A salient pole rotor would be inappropriate, since the centrifugal force acting on the poles would be very large. For example, the force on a 1 kg mass on the edge of a 1 m rotor, rotating at 3000 rev/min, would be about 50 kN. In the cylindrical rotor the field windings are contained in longitudinal slots machined into the outer periphery of the rotor core. Each slot contains a number of conductors which are securely fastened in by means of wedges. The basic construction is illustrated in Fig. 5.10. In addition to its great mechanical strength, the cylindrical rotor offers far less wind resistance (thus reducing the machine losses) and tends to produce a better flux pattern. The latter effect results in a more sinusoidal waveform than is generally available from a salient pole rotor.

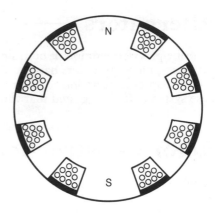

5.7 Alternator emf Equation

Let z = number of stator conductors per phase
Φ = useful flux per pole, in Wb
p = number of *pairs* of poles
n = rotor speed, in rev/s

flux linking 1 conductor in 1 rev. = $2\Phi p$ weber
flux linking 1 conductor in 1 sec. = $2n\Phi p$ weber/second

and since a rate of change of flux of 1 Wb/s induces an emf of 1 volt, then:

average emf generated in 1 conductor = $2n\Phi p$ volt
average emf generated per phase = $2p\Phi zn$ volt

assuming a sinusoidal waveform (form factor of 1.11)

then r.m.s. emf generated per phase = $2.22p\Phi zn$ volt

but frequency $f = np$ hertz, so the r.m.s. emf generated per phase is given by the equation:

$$E \text{ per phase} = 2.22\Phi zf \text{ volt} \qquad (5.6)$$

■ Worked Example 5.3 ■

 A 6-pole, single-phase alternator is driven at 1000 rev/min. There are 36 stator slots each containing 16 conductors. If the useful flux/pole is 0.04 Wb, calculate the frequency and r.m.s. value of the generated emf.

 ———————————————————

$p = 3$; $\Phi = 0.04$ Wb; $n = 1000/60$ rev/s; $z = 36 \times 16$

$$f = np \text{ hertz} = \frac{1000}{60} \times 3 = 50 \text{ Hz } \textbf{Ans}$$

$E = 2.22\Phi zf \text{ volt} = 2.22 \times 0.04 \times 36 \times 16 \times 50$

hence, $E = 2.557$ kV **Ans**

■ Worked Example 5.4 ■

 A three-phase, 50 Hz alternator has a star-connected stator winding. There are 93 stator slots, each containing 4 conductors. If the useful flux/pole is 0.348 Wb, calculate the alternator line voltage.

A

$$f = 50 \text{ Hz}; \Phi = 0.348 \text{ Wb}; z = 4 \times 93/3 = 124$$

$$E_{ph} = 2.22\Phi zf \text{ volt} = 2.22 \times 0.348 \times 124 \times 50$$

$$= 4.79 \text{ kV}$$

$$E_L = \sqrt{3} \times E_{ph} = \sqrt{3} \times 4790$$

therefore, $E_L = 8.3 \text{ kV}$ **Ans**

5.8 Alternator Losses

The input power to the alternator is in two parts; the mechanical driving power to the rotor, ωT watt (or $2\pi n T$), and the d.c. electrical power $V_r I_r$ watt which provides the excitation current for the field winding. The difference between the total input power and the electrical power output accounts for the losses in the machine. These losses consist of the iron, friction and windage losses, and the copper losses due to the resistance of the stator winding. The iron losses are due to hysteresis and eddy current losses in the stator core as the rotor field sweeps past it. There will also be mechanical losses due to friction in the bearings and sliprings, together with the air friction (windage) of the rotor. These losses are best illustrated in the form of a power flow diagram as shown in Fig. 5.11.

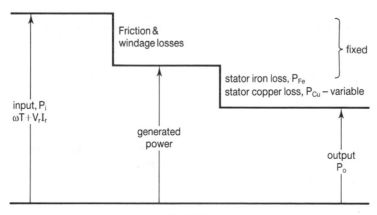

Fig 5.11

The iron, friction and windage losses are sensibly constant, but the stator copper loss will naturally vary according to the amount of output current drawn by the load. Like the transformer, the alternator will operate at its maximum efficiency when the variable losses equal the fixed losses.

■ Worked Example 5.5 ■

Q A 40 kVA, single-phase alternator supplies a full-load current of 100 A at a power factor of 0.85. Under this condition the iron, friction and windage loss is 1500 W, the stator copper loss is 4155 W, and the rotor field is supplied with 5 A d.c. at a p.d. of 100 V. Determine (a) the full-load efficiency, and (b) the input driving torque, if the speed is 3000 rev/min.

A ————————————————————————————

Output = 40 kVA; cos ϕ = 0.85; P_{Fe} = 1500 W;

n = 3000/60 = 50 rev/s; P_{Cu} = 4155 W;
I_r = 5 A; V_r = 100 V

(a) output, P_o = kVA cos ϕ watt = 40 × 0.85
          = 34 kW

total losses = P_{Fe} + P_{Cu} watt
          = 1500 + 4155 = 5655 W

input, P_i = P_o + losses = 34000 + 5655
          = 39655 W

$$\eta = \frac{P_o}{P_i} \times 100\% = \frac{34000}{39655} \times 100$$

hence, η = 85.74% **Ans**

(b) $P_i = \omega T + V_r I_r$, watt

hence $T = \dfrac{P_i - V_r I_r}{\omega}$; where $\omega = 2\pi n$ rad/s

$$= \frac{39655 - 500}{100\pi}$$

therefore, T = 124.6 Nm **Ans**

5.9 The Production of a Rotating Magnetic Field from a Polyphase Supply ————————————————

In order to achieve the motor effect, there has to be an interaction between two magnetic fields. In the case of a.c. motors the stationary stator windings are required to produce a rotating flux pattern that will interact with the flux produced by the rotor winding. When a polyphase supply is connected to an appropriate polyphase winding such a rotating field results, as explained below.

■ Three-phase rotating field ■

Consider a three-phase supply connected to the stator windings of a three-phase machine. For simplicity of explanation consider each phase of the winding to be represented by a single-turn winding, as illustrated in Fig. 5.12. Also shown in this diagram is the waveform of the supply connected to the stator. The diagram of the stator has been repeated at 30° intervals in order to show the directions of the stator currents at these instants of time. These current directions have been deduced by using the following convention.

When the relevant waveform is positive, the stator current will be from R to R′, Y to Y′, or B to B′, as appropriate. Similarly, when each waveform is negative, then the relevant current directions will be reversed, i.e. R′ to R etc. It may be seen that the stator flux will make one complete revolution during one complete cycle of the supply. Thus the field will rotate at the same angular velocity as the supply frequency. For example, if the supply frequency is 50 Hz, then the stator field will rotate at 50 rev/s, or 100π rad/s. Another point to note is that the *amplitude* of this rotating field is *constant*, with the result that a three-phase motor will provide a smooth torque output. Reference to these facts was made in Chapter 3, where the advantages of three-phase systems compared with single-phase were stated.

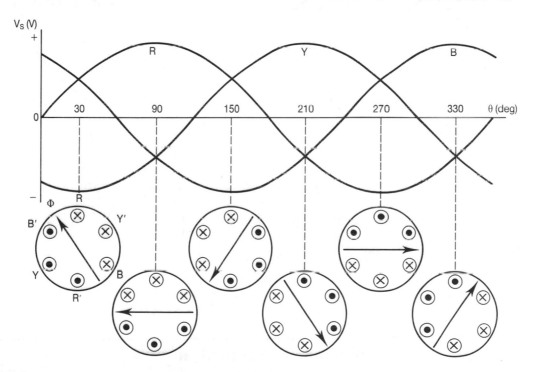

Fig 5.12

■ Two-phase rotating field ■

A two-phase supply consists of two sinewaves that are 90° out of phase with each other. If such a supply is connected to two stator windings that are mutually displaced by 90°, then a rotating magnetic field is produced. This is illustrated in Fig. 5.13. A convention similar to that used for the three-phase system applies.

From Fig. 5.13 it may be seen that the rotating flux produced by the stator windings completes one revolution in one complete cycle of the supply. However, the amplitude of this flux is not constant, since at certain points only one winding is carrying current. For this reason the torque produced by a two-phase motor is not as smooth as that from a three-phase machine.

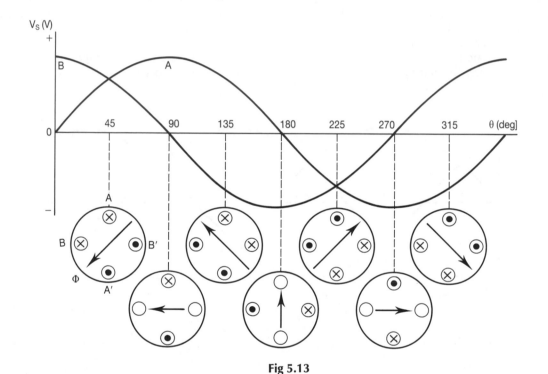

Fig 5.13

■ Single-phase field ■

When a single-phase supply is connected to a single stator winding the flux produced will be pulsating, i.e. it will be sinusoidal. Thus the flux will change its polarity each half-cycle, and its amplitude will be continuously varying. However, such an alternating flux may be considered as consisting of two equal but contra-rotating flux phasors, each of amplitude equal to half the amplitude of the actual flux. The sequence of phasor diagrams shown in Fig. 5.14 illustrates the principle. It may be seen that, at any instant of time, the phasor sum of the two contra-rotating phasors (P_1 and P_2) is equal to the instantaneous value of the flux Φ. Due to the pulsating nature of the flux the torque produced by a single-phase motor is also pulsating. In addition, a single-phase motor is not 'self-starting', as are the polyphase machines. This affect, and the means of overcoming the problem is covered later in this chapter.

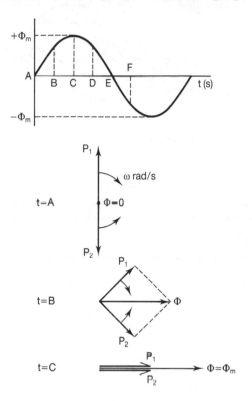

Fig 5.14

5.10 Three-phase Induction Motor

The stator of this machine is identical in construction to that of a three-phase alternator. However, it is usual to bring the six ends of the three windings out to terminals mounted on the motor casing, enabling either star or delta connection, as required. The rotor construction may be of one of two types: a cage rotor or a wound rotor.

■ Cage rotor ■

This consists of aluminium bars contained in the slots of a cylindrical laminated steel core. The ends of the bars are connected together by aluminium shorting rings at each end of the rotor. In small to medium-sized machines these shorting rings are often cast with fins. These fins act as fan blades to circulate a cooling stream of air over the rotor. Figure 5.15 shows the cage construction, but the steel core has been omitted for the sake of clarity. There are no electrical connections to this form of rotor, and this type of machine is sometimes referred to as a brushless a.c. motor.

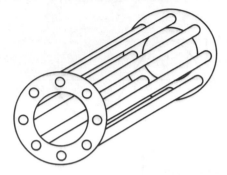

Fig 5.15

■ Wound rotor ■

This also consists of a laminated steel core, but the slots around its periphery contain a three-phase winding that may be connected either in star or delta configuration. In some large machines the three ends of the rotor winding may be connected to slip-rings. This enables external connections to be made for starting purposes, which is explained later.

■ Operating principle ■

This is best understood by considering a single rotor bar on a cage rotor, with the stator field sweeping past it. This is illustrated in Fig. 5.16. The rotating stator field 'cuts' the bar, thus inducing an emf in it. Since the cage rotor forms a complete electrical circuit, then the induced emf causes current to flow, in the direction shown in Fig. 5.16. This direction of induced current may be confirmed by applying Fleming's right-hand rule. However, bear in mind that when this rule is applied it is assumed that the flux is stationary and the conductor is moving. Thus, in Fig. 5.16, the *relative* direction of motion of the conductor with respect to the flux is *anticlockwise*. The circulating rotor current produces its own magnetic field. The interaction between this and the stator field exerts a force on the rotor which will cause it to rotate in the same direction as the stator field. In fact there

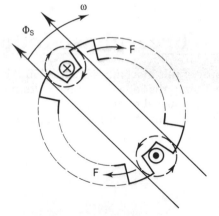

Fig 5.16

will be a torque since the rotor bar diagonally opposite will have an opposite polarity current induced, as shown. As the flux pattern will tend to concentrate in the iron circuit, the torque is exerted on the rotor 'teeth', rather than on the rotor bars. The rotor must turn at some speed less than that of the stator field, otherwise there would be no 'cutting' action between this flux and the rotor bars; no induced emf and current; no interaction of fluxes; and hence no torque. The speed of rotation of the stator field is referred to as the synchronous speed. Thus the machine must always run at less than synchronous speed. However, when the machine is running light (no mechanical load connected) the rotor speed is very near to synchronous speed.

■ Slip ■

The difference between the rotor speed (n_r or w_r) and the synchronous speed (n_s or w_s) is known as the slip (s). This is usually expressed as a percentage, as follows:

$$\text{slip, } s = \frac{n_s - n_r}{n_s} \times 100\% \qquad (5.7)$$

It has been stated that on no-load n_r is very nearly equal to n_s, so the slip is small (maybe 0.1% or 0.2%). However, as the mechanical loading is increased (hence more torque required), the rotor speed falls. This is to be expected, since increased torque requires a stronger rotor flux. This effect can only be achieved by an increase of the 'cutting' action between the rotor and the stator field. This is further confirmed when considering the power flow diagram (Fig. 5.17). In order to provide more power output, the 'generated' power, $\sqrt{3}E_r I_r \cos \phi_r$ watt must be increased proportionately. The required increase in this

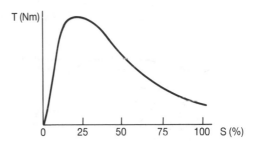

Fig 5.18

power can only result from a larger induced emf, E_r. A typical torque/slip characteristic is shown in Fig. 5.18.

The induction motor is the most commonly used a.c. motor, due to its simplicity of construction (cage rotor type) and its good speed/torque characteristics. Provided that it is not excessively loaded, the variation of speed is not too great, a full-load slip of about 6% being fairly typical.

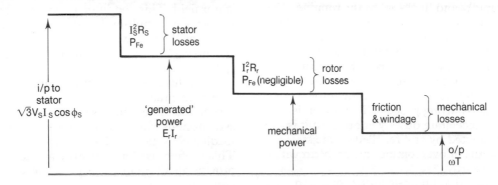

Fig 5.17

5.11 Three-phase Synchronous Motor

The construction of this machine may be identical to that of a three-phase alternator. Indeed, if such an alternator had its stator connected to a suitable three-phase supply, and its rotor field winding was supplied with d.c. current, then the machine would operate as a synchronous motor.

■ Operating principle ■

To illustrate the basic principle consider a three-phase stator winding connected to a supply, but let the rotor be a small bar magnet mounted on a shaft. This is illustrated in Fig. 5.19. As the stator

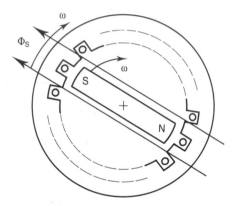

Fig 5.19

field sweeps past the magnet, the North pole of the stator field will attract the South pole of the magnet. Similarly, the diametrically opposite poles will also experience a force of attraction. Provided that the bar magnet has very little inertia it will accelerate rapidly and 'lock-on' to the rotating stator field. Thus our 'rotor' will continue to rotate at synchronous speed. This starting process will apply only when the machine has no mechanical load, and only if the rotor is very light, as just described. In a practical machine having the normal wound rotor, the inertia of the rotor inhibits this self-starting process. The reason is that it cannot accelerate quickly enough, so it is alternately attracted by and repelled by the stator field. The means of overcoming this problem will be described shortly.

Let us assume that on no-load the rotor and

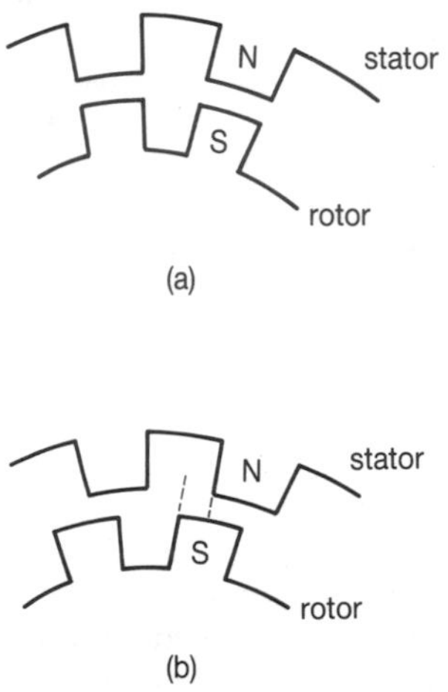

Fig 5.20

stator poles are perfectly aligned, as shown in Fig. 5.20(a). As the mechanical load is increased, this attempts to slow the rotor. The effect is to cause the rotor poles to move slightly out of perfect alignment as shown in Fig. 5.20(b). If the load is increased sufficiently, this misalignment reaches the point where the magnetic attraction is no longer capable of keeping the rotor locked to the stator field. This condition occurs at what is known as the 'pull-out' torque. The result is that the rotor slows down and stops. Thus the synchronous motor can run *only* at synchronous speed.

Since this motor can operate only at one speed it is used in situations where a constant speed drive is required. The other useful feature of the synchronous motor is that, under appropriate loading conditions, it has a *leading* power factor. This is often used in industrial applications to provide an alternative means of power factor correction, instead of power factor correction capacitors. When used in this way the machine may be referred to as a *synchronous capacitor*. The added advantage is that the machine can also be utilised to perform other useful work at the same time. For example, it could be used to provide a constant speed drive for some process, such as driving ventilation fans. The power flow diagram is shown in Fig. 5.21.

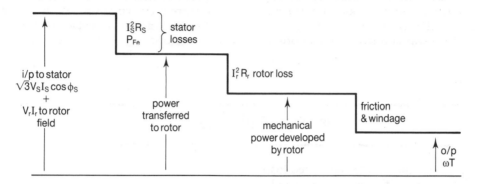

Fig 5.21

5.12 Starting Methods for Three-phase Motors

For small motors (up to about 500 W) of both types, direct connection to the supply is possible. However, for large machines direct on-line starting is avoided. In the case of the synchronous motor self-starting is not possible due to rotor inertia, and in the case of the induction motor damage to both the machine and supply could result.

■ Synchronous motor starting ■

The simplest way to achieve this is to drive the rotor initially by means of an induction motor mounted on the same shaft. Under this condition the induction motor will be running on a very light load, so the slip will be small. Once the induction motor has reached this speed the supply to the synchronous motor stator is connected. The rotor of the synchronous motor will now be able to 'pull-in', 'lock-on' and continue running at synchronous speed. The supply to the induction motor can now be removed. With very large machines there may also be a clutch arrangement between the two motors. In this case, once the synchronous motor is running, the induction motor can also be mechanically disengaged.

■ Induction motor starting ■

Firstly let us consider the affect of connecting a large motor directly to the supply. At that instant the rotor would be stationary, so the slip would be 100%. The resulting 'cutting' action of the stator field would be very great. Since the resistance of the rotor will be very small (especially true for a cage rotor), then an excessively large emf and current would be induced in it. Now, an induction motor may be considered as a rotating transformer, whereby the primary winding is in the stator, and the rotor forms the secondary. Thus direct on-line starting would be akin to putting a short-circuit across the secondary of the transformer whilst it was connected to its normally rated supply. Under this condition the motor stator would draw an excessively large current from the supply. This could cause the stator windings to burn out and/or cause damage to the supply switchgear. Thus some means of reducing the current drawn at start-up is required. There are several methods used, two of which will now be described.

■ Star-Delta starter ■

In Chapter 2 we found that a star-connected load dissipates only one third of the power compared to when it is connected in delta. This fact is utilised in this form of starter. The stator of the induction motor, in this case, would be delta-connected under normal operating conditions. However, when starting the machine, these windings are connected in star configuration. Thus the input power and hence current drawn from the supply is limited. The six ends of the stator winding are brought out to terminals, which are then connected to a switching arrangement as shown in Fig. 5.22. Studying this diagram will show how the switchgear connects the winding firstly in star. Once the machine speed has built up, the stator winding is switched over to the delta connection, enabling the machine to develop its normal operating power. This method would have to be used for a cage rotor machine, but may also be used for the wound rotor type.

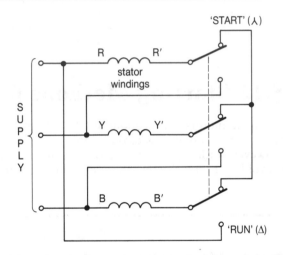

▪ Rotor resistance ▪

For a wound rotor machine, an alternative method is to insert extra resistance into the rotor winding during the start-up period. Once the machine speed has stabilised this additional resistance is reduced to zero. To enable this process the ends of the rotor have to be connected to the external starter via slip-rings. The basic arrangement is illustrated in Fig. 5.23. For obvious reasons this method is not applicable to the cage rotor machine.

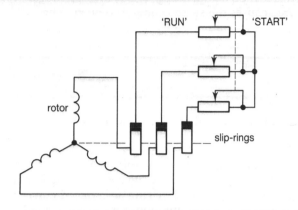

Fig 5.23

5.13 Single-phase Motors

Although two-phase motors exist, they are used only in some specialist applications (such as servo motors) and will not be described here. However, the principle of a two-phase rotating field is utilised in some single-phase motors. The problems associated with a single-phase supply connected to a single stator winding were outlined in Section 5.9. The principal problem is that a single-phase motor is inherently not self-starting. Single-phase motors may be either synchronous or induction machines, and are generally small motors. They find their main usage in low-power applications and where a three-phase supply is not readily available. Examples are motors for mains operated electric clocks, office machinery, and some household appliances such as turntable drive motors for record players. Due to the simplicity of the cage rotor construction, the vast majority of single-phase motors are induction machines of this type.

▪ Split-phase motors ▪

These machines have two stator windings, mounted at 90° to each other. One of these, the 'start' winding (identified as S in Fig. 5.24) has a capacitor connected in series with it. Since the current through a capacitor leads the p.d. across it by 90°, then the current through this winding will be approximately 90° out of phase with respect to that flowing through the main winding, M. We therefore have the same effect as a two-phase supply connected to a two-phase stator winding. The result is that a rotating field is produced, and the machine is self-starting. A variation on this

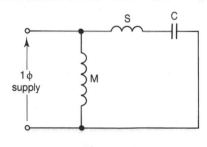

Fig 5.24

arrangement is to include a centrifugal switch, mounted on the rotor shaft, with its contact in series with the start winding. Once the machine has reached its operating speed, the centrifugal force causes the switch contact to open. The machine then continues to run on the main winding, 'following' the appropriate rotating flux phasor component, P_1 or P_2, described in Fig. 5.14. A typical application (without the centrifugal switch) is a central heating pump motor.

■ Shaded pole motor ■

Each pole piece of this machine has a heavy copper ring embedded in one half, as illustrated in Fig. 5.25. Since a.c. is passed through the stator field windings, the flux is varying sinusoidally. The shading ring will therefore have a substantial eddy current induced into it. Being an induced current it obeys Lenz's law, and opposes the change of flux that induced it. The result is that the change of flux in one half of the pole piece is delayed compared with that in the other half. This is equivalent to the production of two flux components that are approximately 90° out of phase, which has the same affect as a two-phase field. The resultant flux pattern will therefore be rotating and this machine also is self-starting.

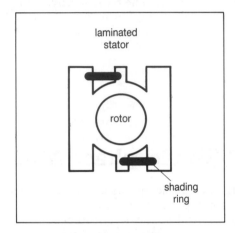

Fig 5.25

■ Universal motor ■

This is actually a series d.c. motor, which can also operate from a single-phase a.c. supply. D.C. motors are fully described in the following chapter, but a brief explanation of this machine now follows. In a series motor the field winding is wound in series with the armature winding (the rotating part). Thus, as the current through the armature reverses each half cycle, so the polarity of the field is reversed. The torque exerted on the armature therefore continues to act in one direction, and the armature is continuously rotated. This machine is the most commonly used single-phase motor. Typical applications are washing machines, vacuum cleaners, hair driers, hot-air hand driers, electric drills, etc. The main reason for the popularity of this motor is that it is simple to vary its speed. It is difficult and expensive to provide speed variation for true a.c. motors.

Assignment Questions

1 A transformer has a voltage ratio of 240 V/12 V. If there are 1500 turns on the primary winding, calculate the number of secondary turns.

2 A single-phase transformer is supplied at 100 V, and a secondary voltage of 440 V is required. If the secondary has 300 turns, determine the number of primary turns required.

3 A transformer supplies a load current of 9.5 A. If the primary to secondary turns ratio is 12:1, calculate the value of the primary current.

4 A step-up transformer has a turns ratio of 2:15. The primary is connected to a 240 V supply, and a 400 Ω resistive load is connected to the secondary. Calculate (a) the secondary voltage, (b) the primary current and (c) the power drawn from the supply.

5 A 1:3 step-up transformer is connected to a 240 V, 50 Hz single-phase supply. If the number of secondary turns is 2700, calculate the peak value of the core flux.

6 A 50 Hz transformer has a core cross-sectional area of 480 cm^2 and a maximum flux density of 1.5 T. The number of turns on the primary and secondary windings are 75 and 250 respectively. Determine the primary and secondary induced emfs.

7 A 660 V/400 V, 25 kVA transformer yielded the following results when tested:

On open-circuit:

$$V_p = 660 \text{ V}; \ I_o = 2 \text{ A}; \ P_1 = 240 \text{ W}$$

On short-circuit:

$$V_p = 50 \text{ V}; \ I_p = 103 \text{ A}; \ P_2 = 500 \text{ W}$$

Sketch the no-load phasor diagram and calculate (a) the no-load power factor, (b) the full-load efficiency, assuming a purely resistive load, and (c) the value of the total losses to ensure maximum efficiency.

8 A 10 kVA transformer has a primary voltage of 415 V, and produces a 100 V output. When connected to this supply it draws a no-load current of 1.2 A at a power factor of 0.3. On full load, and supplying a purely resistive load, the copper losses are 350 W. Calculate (a) the full-load efficiency, (b) the copper loss that would result in maximum efficiency, and (c) the value of maximum efficiency. Sketch the no-load phasor diagram.

9 A 50 Hz alternator has 4 poles, and generates an emf of 800 V. If the useful flux per pole is 35 mWb, calculate (a) the number of conductors/phase, and (b) the speed at which it is driven.

10 A single-phase alternator generates an emf of 250 V at 50 Hz. There are 90 stator slots, two-thirds of which contain 4 conductors each. If it is driven at 1000 rev/min, calculate (a) the number of rotor poles, and (b) the useful flux/pole.

11 A three-phase, two pole, star-connected alternator is driven at 3000 rev/min. The stator has a total of 180 slots, each containing 5 conductors. If the useful flux/pole is 0.04 Wb, calculate (a) the generated emf per phase, (b) the line voltage, and (c) the output power when supplying a balanced load with a current of 15 A at 0.9 power factor.

12 With the aid of a power flow diagram, explain the losses of an alternator, identifying those losses that are fixed and those that are variable.

A 15 kVA single-phase alternator supplies a load with 20 A, at a power factor of 0.78 lagging. The rotor is connected to a 50 V d.c. supply, from which it draws a current of 2 A. Under these conditions the fixed losses are 1.1 kW, and the machine has an efficiency of 80%. Determine (a) the mechanical power input, and (b) the stator copper losses.

13 (a) Explain the advantages of three-phase a.c. motors compared with single-phase motors.

(b) With the aid of a simple sketch, explain how torque is developed on the rotor of an induction motor.

(c) Explain the term 'slip', and state why this must occur in an induction motor.

14 Explain one method of starting a large synchronous motor, and one method for a large three-phase induction motor. Include in your answer the reasons why these starting methods are required.

Suggested Practical Assignments

Assignment 1

To determine the full-load efficiency for a single-phase power transformer.

Apparatus:
1 × single-phase power transformer
1 × wattmeter
1 × voltmeter
1 × variac
2 × ammeters (one of which must be capable of measuring the full-load secondary current)

Method:

1 Carry out the open and short-circuit test on the transformer. These techniques are outlined in Section 5.3.

   IMPORTANT NOTE: When conducting the short-circuit test it is *essential* that the variac is set to *zero volts* output, and this voltage is then increased *very carefully* until the ammeter connected across the secondary indicates the rated full-load secondary current.

2 Tabulate all meter readings from the two tests, and hence calculate the full-load efficiency of the transformer.

Assignment 2

To carry out a load test on a fractional horsepower three-phase induction motor.

Apparatus:
1 × 'low-voltage' 3-phase induction motor
1 × low-voltage 3-phase supply
2 × wattmeters
1 × tachometer
1 × brake test rig

Method:

1 Adjust the belt on the brake test rig so that it is just making contact with the brake drum attached to the motor shaft, and check that the spring balance on this rig indicates zero.

2 Connect the stator windings and the two wattmeters to the switched three-phase supply, as indicated in Fig. 5.26.

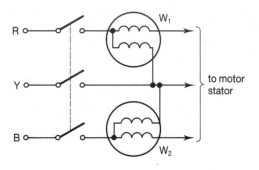

Fig 5.26

3 Connect the supply and note the no-load speed of the motor by means of the tachometer.

4 Increase the mechanical loading on the motor, in steps, by increasing the brake belt tension. Note and tabulate the spring balance reading, the wattmeter readings and motor speed at each step.

5 Continue increasing the mechanical loading, as detailed in paragraph 4 above, until the machine's full rated power is indicated by the sum of the two wattmeter readings.

6 Calculate and tabulate the motor torque, slip, and output power for all loadings applied, using the following equations:

motor torque, $T = Fr$ newton metre

where F = spring balance reading, in newton
and r = radius of brake pulley, in metres

$$P_o = wT \text{ watt}$$

where $\omega = 2\pi n/60$ radian/second, and n = speed in rev/minute

7 Calculate the full-load efficiency of the motor.

8 Plot graphs of torque versus slip, and speed versus output power.

9 Complete a full assignment report.

■ 6 D.C. Machines ■

This chapter covers the operating principles of d.c. generators and motors, their characteristics and applications. On completion you should be able to:

1] Understand and explain generator/motor duality.

2] Appreciate the need for a commutator, the problems associated with the process of commutation and methods of overcoming these. Explain the rectifying action of a commutator.

3] Appreciate the problem of armature reaction, and methods used to reduce its affect. Describe the construction of d.c. machines.

4] Apply the emf equation and appreciate the affect on this equation of different types of armature winding.

5] Identify the different types of d.c. generator, and describe their characteristics. Carry out practical tests to compare the practical and theoretical characteristics.

6] Deduce the relationships between emf, torque, speed and current, and use these to deduce the characteristics of d.c. motors.

7] Explain the need for, and describe the operation of a d.c. motor starter.

8] List and explain the losses of d.c. machines. Carry out calculations involving these losses in order to predict the efficiency of d.c. machines.

9] Explain different methods of simple speed control for d.c. motors.

10] Explain the principles of simple stepper motors.

6.1 Motor/generator Duality

An electric motor is a rotating machine which converts an electrical input power into a mechanical power output. A generator converts a mechanical power input into an electrical power output. Since one process is the converse of the other, a motor may be made to operate as a generator, and vice versa. This duality of function is not confined to d.c. machines. It was mentioned in Chapter 5 that an alternator can be made to operate as a synchronous a.c. motor, and vice versa.

To demonstrate the conversion process involved, let us reconsider two simple cases that were met when dealing with electromagnetic induction, in Volume 1.

Consider a conductor being moved at constant velocity, through a magnetic field of density B tesla, by some externally applied force F newton. This situation is illustrated in Fig. 6.1.

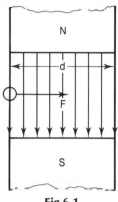

Fig 6.1

Work done in moving the conductor,

$$W = Fd \text{ newton metre}$$

mechanical power input, $P_1 = \dfrac{W}{t}$ watt

$$\text{so, } P_1 = \dfrac{Fd}{t} \text{ watt}$$

and since d/t is the velocity, v at which the conductor is moved, then

$$P_1 = Fv \text{ watt } \dots\dots\dots \text{ [1]}$$

However, when the conductor is moved, an emf will be induced into it. Provided that the conductor forms part of a closed circuit, then the resulting current flow will be as shown in Fig. 6.2. This induced current, i, produces its own magnetic field, which reacts with the main field, producing a reaction force, F_r, in direct opposition to the applied force, F.

$$\text{Now, } F_r = Bil \text{ newton}$$

Assuming no frictional or other losses, then the applied force has only to overcome the reaction force, such that:

$$F = F_r = Bil \text{ newton}$$

so equ.[1] becomes $P_1 = Bilv$ watt ... [2]

Also, induced emf, $e = Blv$ volt

so generated power, $P_2 = ei$ watt

therefore, $P_2 = Bilv$ watt ... [3]

Since [3] = [2], then the electrical power generated is equal to the mechanical power input (assuming no losses). Now consider the conductor returned to its original starting position. Let an external source of emf, e volt pass a current of i ampere through the conductor. Provided that the direction of this current is opposite to that shown in Fig. 6.2, then the conductor will experience a

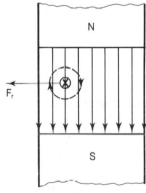

Fig 6.2

force that will propel it across the field. In this case, the same basic arrangement exhibits the motor effect, since the electrical input power is converted into mechanical power.

Although the above examples involve linear movement of the conductor, exactly the same principles apply to a rotating machine.

6.2 The Generation of d.c. Voltage

We have seen already that, if a single-loop coil is rotated between a pair of magnetic poles, then an *alternating* emf is induced into it. This is the principle of a simple form of alternator. Of course, this a.c. output could be converted to d.c. by employing a rectifier circuit. Indeed, that is exactly what is done with vehicle electrical systems. However, in order to have a truly d.c. machine, this rectification process needs to be automatically accomplished within the machine itself. This process is achieved by means of a commutator, the principle and action of which will now be described.

Consider a simple loop coil, the two ends of which are connected to a single 'split' slip-ring, as illustrated in Fig. 6.3. Each half of this slip-ring is

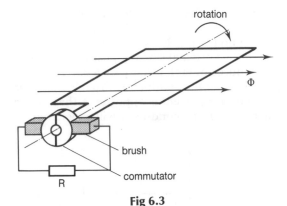

Fig 6.3

insulated from the other half, and also from the shaft on which it is mounted. This arrangement forms a simple commutator, where the connections to the external circuit are via a pair of carbon brushes. The rectifying action is demonstrated in the series of diagrams of Fig. 6.4. In these diagrams, one side of the coil and its associated commutator segment are identified by a thickened line edge. For the sake of clarity, the physical connection of each end of the coil, to its associated commutator segment, is not shown. Fig. 6.4(a) shows the instant when maximum emf is induced in the coil. The current directions have been determined by applying Fleming's right-hand rule. At this instant current will be fed out from the coil, through the external circuit from right to left, and back into the other side of the coil. As the coil continues to rotate from this position, the value of induced emf and current will decrease. Figure 6.4(b) shows the instant when the brushes short-circuit the two commutator segments. However, the induced emf is also zero at this instant, so no current flows through the external circuit. Further rotation of the coil results in an increasing emf, but of the opposite polarity to that

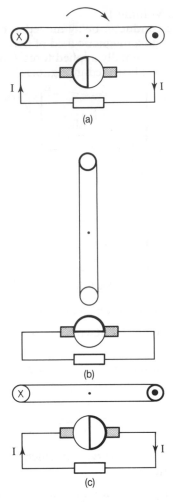

Fig 6.4

induced before. Figure 6.4(c) shows the instant when the emf has reached its next maximum. Although the generated emf is now reversed, the current through the external circuit will be in the same direction as before. The load current will therefore be a series of half-sinewave pulses, of the same polarity. Thus the commutator is providing a d.c. output to the load, whereas the armature generated emf is alternating.

A single-turn coil will generate only a very small emf. An increased amplitude of the emf may be achieved by using a multi-turn coil. The resulting output voltage waveform is shown in Fig. 6.5. Although this emf is unidirectional, and may have a satisfactory amplitude, it is not a satisfactory d.c.

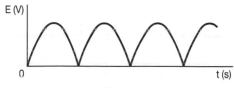

Fig 6.5

waveform. The problem is that we have a concentrated winding. In a practical machine the armature has a number of multi-turn coils. These are distributed evenly in slots around the periphery of a laminated steel core. Each multiturn coil has its own pair of slots, and the two ends are connected to its own pair of commutator segments. Figure 6.6 shows the armature construction (before the coils have been inserted). The riser is the section of the commutator to which the ends of the coils are soldered. Due to the distribution of the coils around the armature, their maximum induced emfs will occur one after the other, i.e. they will be out of phase with each other. Figure 6.7 illustrates this, but for simplicity, only three coils have been considered. Nevertheless, the effect on the resultant machine output voltage is apparent, and is shown by the thick line along the peaks of the waveform. With a

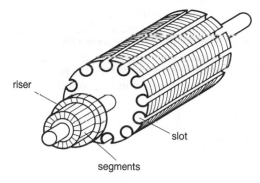

Fig 6.6

large number of armature coils the ripple on the resultant waveform will be negligible, and a smooth d.c. output is produced.

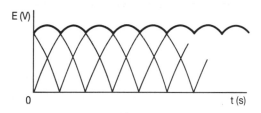

Fig 6.7

6.3 The Commutation Process

Commutation is the process whereby the current in an armature coil is reduced to zero, reversed, and increased in the opposite direction. The problem is that this process needs to be carried out in the very short instant of time that the coil's armature segments are short-circuited by one of the carbon brushes. This problem is further aggravated by the inductance of the coils, since this will result in a delay in both growth and decay of current (refer back to Chapter 4, dealing with d.c. transients). The commutation process is illustrated in Figs. 6.8 to 6.10 inclusive. Each of these diagrams represents a particular instant of time, as the commutator segments pass over a brush.

In Fig. 6.8, current from opposite halves of the armature winding arrive at the brush via coils 1 and 2 and segment A. Figure 6.9 shows the instant that segments A and B are short-circuited by the

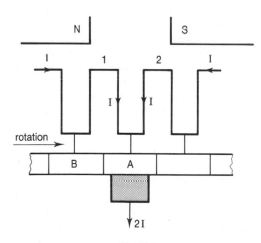

Fig 6.8

brush. At this instant coil 1 was carrying I amps, but as it is now short-circuited, the current needs to be zero. Thus the current in coil 1 is *trying* to reduce to zero instantaneously. This will induce a back emf, known as the reactance voltage E_r, into coil 1. This reactance voltage therefore opposes the decay of current. For illustrative purposes only, it has been assumed in Fig. 6.9 that the

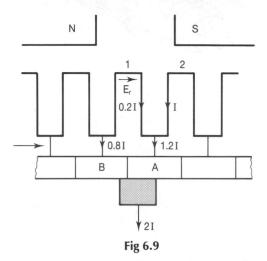

Fig 6.9

current has succeeded in decreasing to 20% of its original value at this time. In Fig. 6.10, segment A has just left contact with the brush. Ideally, under this condition, the full current of I amps should be flowing through coil 1, from the right-hand side of the armature. However, the reactance voltage, although decreased, will still be trying to maintain current flow in the opposite direction. For illustrative purposes it has been assumed that the current through coil 1 has succeeded in reversing to only 90% of its full value. Since Kirchhoff's current law must be maintained, the remaining 10% of the current from this side of the armature must reach the brush. The only way in which this can occur is for this proportion of the current to

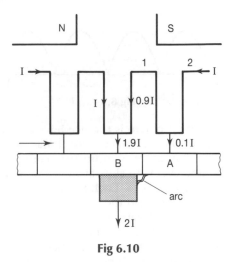

Fig 6.10

form an arc from segment A to the brush. This arcing will cause burning and pitting to both the commutator segments and the trailing edges of the brushes. This is obviously undesirable, and some means of reducing the sparking at the brushes is needed.

The key to the solution of the problem lies in the reactance voltage. If the short-circuited coil can have an emf induced in it that is equal and opposite to the reactance voltage, then there will be no opposition to the reversal of the coil current. One simple solution would be to advance the brushes, in the direction of rotation, so that the short-circuited coil comes under the influence of the next main pole ahead. Although this solution will reduce the sparking at the brushes, it also has another adverse effect. This will be described under the heading of armature reaction, in the next section of this chapter. However, for small motors and generators, the repositioning of the brushes is sometimes adopted.

6.4 Armature Reaction

Figure 6.11 represents an armature rotating between two poles. The direction of induced emfs will be as shown, but as the brushes are not connected to any external circuit, no armature current will flow. The magnetic neutral axis (MNA) is the axis along which zero emf is

induced, and in this case it coincides with the geometric neutral axis (GNA).

If a load is now connected to the brushes, armature current will flow. This will result in a second magnetic field, the armature reaction flux Φ_A, as shown in Fig. 6.12. The interaction of this

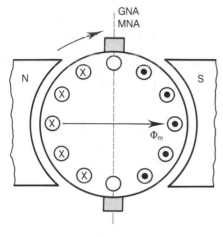

Fig 6.11

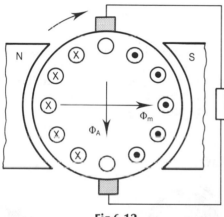

Fig 6.12

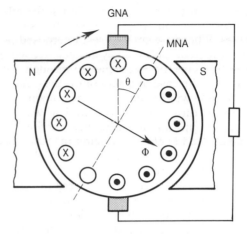

Fig 6.13

to advance the brushes, through the angle θ, on to the new MNA. This solution would also provide the bonus of reducing the reactance voltage, thus reducing arcing due to commutation. However, there are two main problems with this solution. Firstly, the angle θ will vary as the load, and hence armature current, varies. Thus the brushes would have to be repositioned each time the load changed. This would not be very practical. Secondly, when the brushes are advanced, a band of armature conductors produce an mmf in direct opposition to the main field. This weakens the main field, and thereby reduces the value of generated emf. This band of demagnetising conductors is shown in Fig. 6.14.

flux with the main field from the poles results in a resultant flux. This resultant is, in effect, a distorted version of the main field, which results in a shift of the MNA through some angle θ degree in the direction of rotation. This effect is shown in Fig. 6.13.

For good commutation the brushes need to lie along the MNA, so that the brushes short-circuit armature coils only when zero emf is induced in them. Due to the armature reaction effect, brushes remaining on the GNA will be short-circuiting coils having emf induced in them. This will have the effect of reducing the overall emf available to the output, in addition to causing further arcing at the brushes. One solution to the problem would be

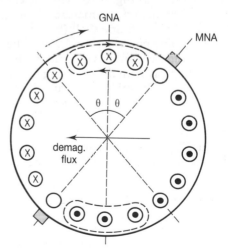

Fig 6.14

For medium to large sized machines, the effects of armature reaction are reduced by the use of interpoles. These are small poles, interposed between the main poles. They are magnetised by the armature current that is passed through their windings. In the case of a generator, the polarity of each interpole will be that of the next main pole ahead, in the direction of rotation. This will have the effect of opposing the armature reaction flux, with the added benefit of reducing the reactance voltage due to commutation. The strength of the interpole flux required to cancel out armature reaction flux may not be that required to cancel the reactance voltage. In practice a compromise is achieved so that a considerable reduction in both is achieved. Note that the strength of the interpole flux will depend directly upon the load, and hence the value of armature current. Thus this flux will automatically compensate for variation of machine load. The affect of and positioning of interpoles is illustrated in Fig. 6.15.

It has already been concluded that a generator will operate as a motor, rotating in the same direction, if it is supplied with an armature current in the opposite direction to the generated current. The interpoles in such a machine will now have the polarity of the nearest main pole behind, with

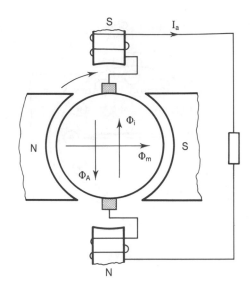

Fig 6.15

respect to the direction of rotation. This is the appropriate condition for a motor in order to overcome the armature reaction effect. Thus the inclusion of interpoles does not inhibit the generator/motor duality.

6.5 Construction of D.C. Machines

The various parts of a small d.c. machine are shown separately in Fig. 6.16, with the exception that neither the field nor armature windings have been included. The frame shell (bottom left) contains the pole pieces, around which the field winding would be wound. One end frame (top left) would simply contain a bearing for the armature shaft. The other end frame (bottom right) contains the brushgear assembly in addition to the other armature shaft bearing. The armature (top right) construction has already been described. The slots are skewed to provide a smooth starting and slow-speed torque.

Fig 6.16

6.6 Types of Armature Winding

There are two main forms of armature winding, known as wave and lap windings. The essential difference between them being the distance between the commutator segments to which the ends of the coils are connected.

▪ Wave winding ▪

Each coil of the winding spans one pole pitch. This means that one side of the coil is directly under one pole when the other side is directly under an opposite polarity pole. However, the two ends of the coil do not terminate at adjacent commutator segments. The arrangement is illustrated in Fig. 6.17. With the wave winding

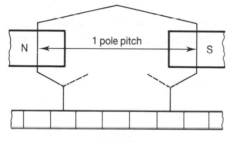

Fig 6.17

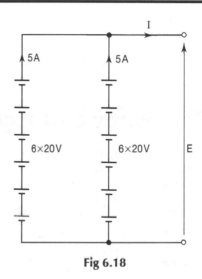

Fig 6.18

there are two parallel paths through the armature, regardless of the number of poles. If we consider each coil of the winding as an individual source of emf, then the machine may be represented by the equivalent circuit of Fig. 6.18. For the sake of explanation, let us assume that each source has an emf of 20 V, and produces a currentof 5 A. It may be seen that the total machine emf is equal to the sum of the emfs of half of the coils connected in series. The total machine current is equal to twice the coil current. In the example sown, the machine emf would be 120 V, and the current would be 10 A.

▪ Lap winding ▪

With this winding each coil would again span one pole pitch, but the two ends would terminate at adjacent commutator segments. This is shown in Fig. 6.19. With this form of winding there are as many parallel paths through the armature as there are poles. If it is a two pole machine, then this will have the same effect as the wave winding.

However, let us consider a four pole machine, and the armature equivalent circuit as shown in Fig. 6.20. Since there are four parallel paths, then the total machine emf will be 60 V, and the machine current will be 20 A.

Comparing these two types of winding, provided that the machine has four (or more) poles, then a

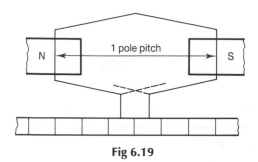

Fig 6.19

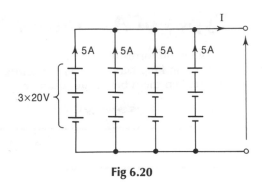

Fig 6.20

wave wound armature provides 'high' voltage, 'low' current output; whereas the lap winding provides 'low' voltage, 'high' current output. Note that the terms 'low' and 'high' are used here only

as relative terms when comparing the two winding types.

6.7 Generator emf Equation

Let z = total number of armature conductors
p = number of pole *pairs*
n = armature speed in rev/second
Φ = useful flux per pole
a = number of parallel paths through armature

Flux cut by one conductor in one rev. = $2p\Phi$ weber
flux cut by one conductor per second = $2p\Phi n$ weber
hence emf generated in one conductor = $2p\Phi n$ volt

since the machine has 'a' parallel armature paths, then the effective number of conductors is z/a, hence

total generated machine emf, $E = \dfrac{2p\Phi zn}{a}$ volt (6.1)

Note: For a lap wound armature, $a = 2p$
and for a wave wound armature, $a = 2$

■ Worked Example 6.1 ■

Q A six-pole armature is wound with 480 conductors. The flux and speed are such that the emf generated in each conductor is 1.5 V, and the current flowing through each conductor is 100 A. Determine the total generated emf and power produced if the winding is (a) wave wound, and (b) lap wound.

A _____

$p = 3$; $z = 480$; $e = 1.5$ V; $i = 100$ A

    (a) When wave wound number of parallel paths = 2, so there will be 240 conductors in each half.

Thus, total emf, $E = 240 \times e$ volt = 360 V **Ans**

and total current, $I = 2 \times i$ amp = 200 A **Ans**

and power output, $P = EI$ watt = 72 kW **Ans**

(b) When lap wound, number of parallel paths = $2p$ = 6, so there will be 480/6 = 80 conductors per path.

Thus, total emf, $E = 80 \times e$ volt = 120 V **Ans**

and total current, $I = 6 \times i$ amp = 600 A **Ans**

and power output, $P = EI$ watt = 72 kW **Ans**

▪ Worked Example 6.2 ▪

 A four-pole armature has 600 lap-connected conductors, and is driven at 1500 rev/min. Calculate the useful flux per pole required to generate an emf of 250 V.

$p = 2$; $z = 600$; $n = 1500/60$ rev/s; $a = 4$

$$E = \frac{2p\Phi zn}{a} \text{ volt}$$

therefore, $\Phi = \dfrac{Ea}{2pzn}$ weber $= \dfrac{250 \times 4 \times 60}{2 \times 2 \times 600 \times 1500}$

hence, $\Phi = 16.67$ mWb **Ans**

6.8 Classification of Generators

D.C. generators are classified according to whether the field winding is electrically connected to the armature winding, and if it is, whether it is connected in parallel with or in series with the armature. The field current may also be referred to as the excitation current. If this current is supplied internally, by the armature, the machine is said to be self-excited. When the field current is supplied from an external d.c. source, the machine is said to be separately excited. The circuit symbol used for the field winding of a d.c. machine is simply the same as that used to represent any other form of winding. The armature is represented by a circle and two 'brushes'. The armature conductors, as such, are not shown.

6.9 Separately excited Generator

The circuit diagram of a separately-excited generator is shown in Fig. 6.21. The rheostat, R_1, is included so that the field excitation current, I_f, can be varied. This diagram also shows the armature being driven at constant speed by some primemover. Since the armature of any generator must be driven, this drive is not normally shown. The load, R_L, being supplied by the generator may

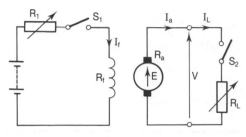

Fig 6.21

be connected or disconnected by switch S_2. The resistance of the armature circuit is represented by R_a.

Consider the generator being driven, with switches S_1 and S_2 both open. Let the armature terminals be monitored by a high impedance voltmeter (a DVM), such that negligible armature current flows. This meter will therefore indicate the generated emf. Despite the fact that there will be zero field current, a small emf would be measured. This emf is due to the small amount of residual magnetism retained in the poles. With switch S_1 now closed, the field current may be increased in discrete steps, and the corresponding values of generated emf noted. If the corresponding graph is plotted, it will be as shown in Fig. 6.22, and is known as the open-circuit

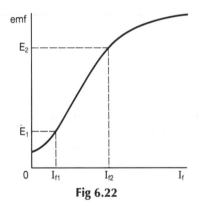

Fig 6.22

characteristic of the machine. It will be seen that the shape of this graph is similar to the magnetisation curve for a magnetic material. This is to be expected, since the emf will be directly proportional to the pole flux (refer back to the emf equation, and consider the terms p, z, a and n to be constants). The 'flattening' of the emf graph indicates the onset of saturation of the machine's magnetic circuit. When the machine is used in

practice, the field current would normally be set to some value within the range indicated by I_{f1} and I_{f2} on the graph. This means that the facility exists to vary the emf between the limits E_1 and E_2 volts, simply by adjusting rheostat R_1.

Let the emf be set to some value E volt, within the range specified above. If the load is now varied, the corresponding values of terminal voltage, V and load current I_L may be measured. Note that with this machine the armature current is the same as the load current. The graph of V versus I_L is known as the output characteristic of the generator, and is shown in Fig. 6.23. The

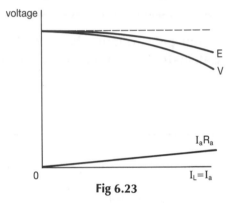

Fig 6.23

terminal p.d. of the machine will be less than the generated emf, by the amount of internal voltage drop due to R_a, such that:

$$V = E - I_a R_a \text{ volt} \tag{6.2}$$

Ideally, the graph of E versus I_L would be a horizontal line. However, the affects of armature reaction cause this graph to 'droop' at the higher values of current. The main advantage of this type of generator is that there is some scope for increasing the generated emf in order to offset the internal voltage drop, $I_a R_a$, as the load is increased. The big disadvantage is the necessity for a separate d.c. supply for the field excitation. The power flow diagram is shown in Fig. 6.24.

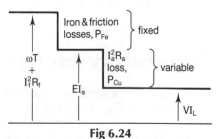

Fig 6.24

6.10 Shunt Generator

This is a self-excited machine, where the field winding is connected in parallel (shunt) with the armature winding. The circuit diagram is shown in Fig. 6.25, and from this it may be seen that the armature has to supply current to both the load and the field, such that:

$$I_a = I_L + I_f \text{ amp} \qquad (6.3)$$

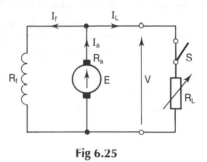

Fig 6.25

▪ Self-excitation process ▪

This process can take place only if there is some residual flux in the poles, and if the resistance of the field circuit is less than some critical value. The open-circuit characteristic is illustrated in Fig. 6.26, where lines OA, OB and OC represent

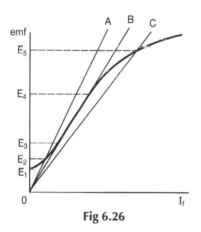

Fig 6.26

different values of shunt field resistance. When the armature is rotated at constant speed, and with switch S open, the initial emf generated will be E_1 volt, due to the residual magnetism. This will cause a small current to circulate through the field winding. This current will cause an increase in the flux, and a consequent increase in emf. However, if the field resistance (represented by OA) is too high, the generated emf will build up only a small amount, to E_2 volt. Thus the machine has failed to

self-excite. Line OB represents the critical value of field resistance, referred to above. With this value of resistance the machine emf would build up to some value within the limits E_3 to E_4 volt. The machine has been able to self-excite. However, the emf is in a highly unstable state, since any slight disturbance, such as a fluctuation in speed, could cause the emf to vary rapidly between the above limits. Thus the field resistance needs to be less than this critical value as represented by OC. Considering this line OC, it may be seen that at all points below E_5 volt, the generated emf is more than that required to merely maintain I_f constant. Hence, more current flows, more flux produced, and more emf is generated. Once the emf reaches E_5 volt, this is sufficient only to maintain I_f constant, so the machine emf stabilises at this value.

The resistance of the field winding, R_f, is constant and of a relatively high value compared with R_a. Typically, I_f will be in the order of 1 A to 10 A, and will remain reasonably constant. The shunt machine is therefore considered to be a constant-flux machine. When switch S is closed, the armature current will increase in order to supply the demanded load current, I_L. Thus $I_a \propto I_L$, and as the load current is increased, so the terminal voltage will fall, according to the equation, $V = E - I_a R_a$ volt. The output characteristic will therefore follow much the same shape as that for the separately excited generator. This condition applies until the machine is providing its rated full-load output. This means that (ignoring any losses) all of the mechanical input power, ωT, is being converted into electrical output power, VI_L watt. If the load should now demand even more current, this will be supplied by the machine, but only at the expense of field current. Thus, part of I_f is diverted to the load. This means that the flux will be reduced; E is reduced; V is reduced; I_L tries to increase to maintain the output power demand; I_f is further

reduced; and so on. The result is that if the generator is overloaded, then it simply stops generating. This effect is shown by the dotted lines in Fig. 6.27, the output characteristics.

The shunt generator is the most commonly used d.c. generator, since it provides a reasonably constant output voltage over its normal operating range. Its other obvious advantage is the fact that it is self-exciting, and therefore requires only some mechanical means of driving the armature. Figure 6.28 shows the power flow diagram.

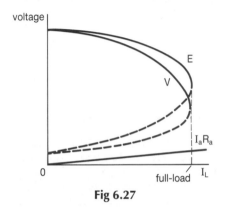

Fig 6.27

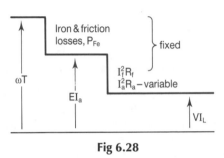

Fig 6.28

6.11 Series Generator

In this machine the field winding is connected in series with the armature winding and the load, as shown in Fig. 6.29. In this case, $I_L = I_a = I_f$, so

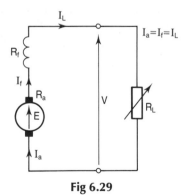

Fig 6.29

this is a variable-flux machine. Since the field winding must be capable of carrying the full-load current (which could be in hundreds of amps for a large machine), it is usually made from a few turns of heavy gauge wire or even copper strip. This also has the advantage of offering a very low resistance.

This generator is a self-excited machine, provide that it is connected to a load when started. Note that a shunt generator will self-excite only when *disconnected* from its load.

When the load on a series generator is increased, the flux produced will increase, in almost direct proportion. The generated emf will therefore increase with the demanded load. The increase of flux, and hence voltage, will continue until the onset of magnetic saturation, as shown in the

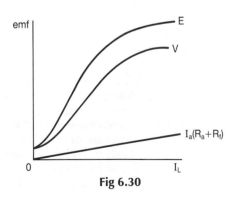

Fig 6.30

output characteristic of Fig. 6.30. The terminal voltage is related to the emf by the equation:

$$V = E - I_a(R_a + R_f) \text{ volt} \qquad (6.4)$$

The variation of terminal voltage with load is not normally a requirement for a generator, so this form of machine is seldom used. However, the rising voltage characteristic of a series-connected field winding is put to good use in the compound machine.

6.12 Compound Generator

This machine contains both shunt and series field windings, wound on to the same poles. The compound generator therefore combines the characteristics of both the shunt and series generators. The affect of the series field winding is designed to be relatively weak compared with that of the shunt field. In addition, the series field winding may be connected so as to either assist (strengthen) the shunt field, or to oppose (weaken) it. The former connection is referred to as *cumulative* compounding, and the latter as *differential* compounding. The circuit diagram for a compound generator is shown in Fig. 6.31. From this diagram it may be seen that the shunt field may be connected on either the armature side or load side of the series field winding. Since the shunt field is the predominant one, it is normally connected directly to the source from which it is supplied. Thus in the case of a generator it is

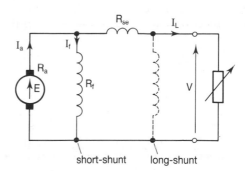

Fig 6.31

connected directly to the armature, and this is called a *short shunt* connection. In the case of a motor, the source of supply is applied to the machine terminals, so the connection shown dotted in the circuit diagram is used. This is referred to as a *long shunt* connection.

■ Cumulative compounding ■

With this connection, rising characteristic of the series field is used to compensate for the gradually falling characteristic of the shunt connection. Depending upon the relative strength of the series field, the machine may be either level compounded or slightly over compounded, as shown in Fig. 6.32. Level compounding is used for applications where excellent voltage regulation is required, i.e. terminal voltage on full-load is the same as that on no-load. If the electrical power output is to be transmitted some considerable distance, then the machine could be over compounded. The increase of terminal voltage with load would then compensate for the increased voltage drop along the transmission lines. If the degree of over-compounding is correctly chosen, then the voltage at the receiving end will remain constant, from no-load to full-load.

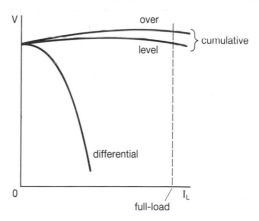

Fig 6.32

▪ Differential compounding ▪

To achieve this, the connections to the series winding would be reversed, so that the current flowing through it will produce a flux in direct opposition to the shunt field flux. The total machine flux will therefore be the *difference* between the two fields. The consequence of this is that, as the load is increased, so the strength of the series flux is increased, but the overall machine flux is *decreased*. The result is the sharply falling voltage characteristic shown in Fig. 6.32. There are only limited applications for this type of characteristic, one of which is the supply to electric arc welding equipment. In this application, a relatively high voltage is required to strike the arc, but a comparatively lower voltage is sufficient to maintain the arc current.

The power flow diagram for a compound generator is shown in Fig. 6.33.

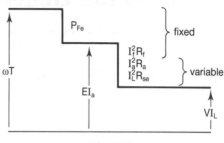

Fig 6.33

▪ Worked Example 6.3 ▪

 Q The resistance of the field winding of a shunt generator is 200 Ω. When the machine is delivering 80 kW the generated emf and terminal voltage are 475 V and 450 V respectively. Calculate (a) the armature resistance, and (b) the value of generated emf when the output is 50 kW, the terminal voltage then being 460 V.

A ──────────────────────────────────

$R_f = 200\ \Omega$; $P_o = 80 \times 10^3$ watt; $V = 450$ V; $E = 475$ V

The circuit diagram is shown in Fig. 6.34. It is always good practice to sketch the appropriate circuit diagram when solving machine problems.

(a)
$$P_o = VI_L \text{ watt; so } I_L = \frac{P_o}{V} \text{ amp}$$

$$\text{therefore } I_L = \frac{80 \times 10^3}{450} = 177.8 \text{ A}$$

$$I_f = \frac{V}{R_f} \text{ amp} = \frac{450}{200} = 2.25 \text{ A}$$

$$I_a = I_L + I_f \text{ amp} = 180.05 \text{ A}$$

$$I_a R_a = E - V \text{ volt} = 475 - 450 = 25 \text{ V}$$

$$\text{therefore } R_a = \frac{25}{180.05} \text{ ohm} = 0.139\ \Omega \text{ Ans}$$

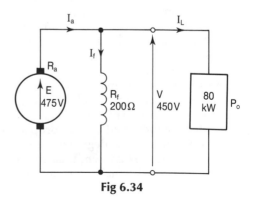

Fig 6.34

(b)   When $P_o = 50 \times 10^3$ W, $V = 460$ V

$$\text{thus } I_L = \frac{50 \times 10^3}{460} = 108.7 \text{ A}$$

$$I_f = \frac{V}{R_f} = \frac{460}{200} = 2.3 \text{ A}$$

hence, $I_a = 108.7 + 2.3 = 111$ A

$E = V + I_a R_a$ volt $= 460 + (111 \times 0.139)$

therefore $E = 475.4$ V **Ans**

Note: Although the load has changed by about 60%, the field current has changed by only about 2.2%. This justifies the statement that a shunt generator is considered to be a constant-flux machine.

▪ Worked Example 6.4 ▪

 A short-shunt compound generator has armature, shunt-field and series-field resistances of **0.75 Ω, 100 Ω and 0.5 Ω respectively. When supplying a load of 5 kW at a terminal voltage of 250 V, calculate the generated emf.**

A

$R_a = 0.75$ Ω; $R_f = 100$ Ω; $R_{Se} = 0.5$ Ω

$P_o = 5 \times 10^3$ W; $V = 250$ V

$$I_L = \frac{P_o}{V} \text{ amp} = \frac{5000}{250} = 20 \text{ A}$$

$V_f = V + I_L R_{Se}$ volt $= 250 + (20 \times 0.5)$

so $V_f = 260$ V

$$I_f = \frac{V_f}{R_f} \text{ amp} = \frac{260}{100} = 2.6 \text{ A}$$

$I_a = I_L + I_f$ amp $= 22.6$ A

$E = V_f + I_a R_a$ volt $= 260 + (22.6 \times 0.75)$

hence $E = 276.95$ V **Ans** (say 277 V)

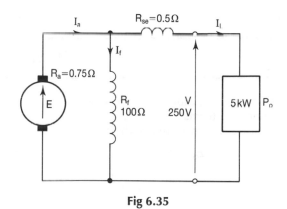

Fig 6.35

6.13 D.C. Motors

All of the d.c. generators so far described could be operated as motors, provided that they were connected to an appropriate d.c. supply. Thus, the effects and problems associated with commutation and armature reaction apply equally to motors.

Similarly the type of armature winding will have the same effect on the emf produced in the armature. When the machine is used as a motor, the armature generated emf is referred to as the back-emf, E_b.

■ Back-emf equation ■

As the armature of a motor rotates, an emf is induced in the armature winding, due to its motion through the magnetic field. The value of this back-emf will be exactly the same as would be generated if the machine was driven as a generator at the same speed. Thus, equation (6.1) could be expressed as follows:

$$E_b = \frac{2p\Phi zn}{a} \text{ volt} \qquad (6.5)$$

The relationship between the back-emf and the supply voltage applied to the motor terminals would be a modified version of equation (6.2), such that:

$$E_b = V - I_a R_a \text{ volt} \qquad (6.6)$$

Note: For a generator, the emf must always be greater than the resulting terminal voltage. It must be borne in mind that for a motor to operate, current must be fed *into* it. Thus the supply voltage, V, applied to the machine terminals must be *greater than* the back-emf, E_b, in order to force current into the machine.

■ Speed equation ■

Considering equation (6.5), for any given machine, the terms $2p$, Φ, a and z are all constants. Thus we can say that:

$$E_b \propto \Phi n$$

$$\text{so, } n \propto \frac{E_b}{\Phi} \text{ or, } \omega \propto \frac{E_b}{\Phi} \qquad (6.7)$$

On examination, this relationship appears to be rather a strange result, since for a given value of back-emf, the *weaker* the flux, the *faster* the resulting speed of rotation. However, this fact is borne out in practice.

■ Torque equation ■

If some of the machine losses are ignored for a moment, then we can say that the mechanical power output is equal to the electrical power 'generated' in the armature, thus:

$$\omega T = E_b I_a$$

$$\text{so, } T = \frac{E_b I_a}{\omega}$$

and substituting for ω, from equation (6.7):

$$T \propto \frac{E_b I_a}{E_b/\Phi}$$

therefore, $T \propto \Phi I_a$ (6.8)

6.14 Shunt Motor

When the machine reaches its normal operating temperature, R_f will remain constant. Since the field winding is connected directly to a fixed supply voltage, V volt, then I_f will be fixed. Thus, the shunt motor (Fig. 6.36) is a constant-flux

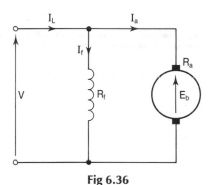

Fig 6.36

machine. Using the speed and torque equations we can say that:

Since $\omega \propto E_b/\Phi$, and Φ is constant; then

$$\omega \propto E_b \dots\dots [1]$$

Similarly, since $T \propto I_a \Phi$; then

$$T \propto I_a \dots\dots [2]$$

As the back-emf will have the same shape graph as that for the generator emf, and using [1] and [2] above the graphs of speed and torque versus current will be as in Fig. 6.37. Note that when the machine is used as a motor, the supply current is identified as I_L. In this case, the subscript 'L'

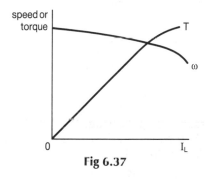

Fig 6.37

represents the word 'line'. Thus I_L identifies the line current drawn from the supply, and I_a is directly proportional to I_L.

Shunt motors are used for applications where a reasonably constant speed is required, between no-load and full-load conditions. The power flow diagram for any d.c. motor is, effectively, the power flow diagram of the corresponding generator, but turning this diagram 'back-to-front'. This is illustrated in Fig. 6.38, which is for a shunt motor. Compare this with Fig. 6.28.

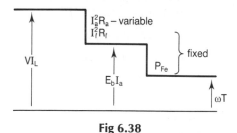

Fig 6.38

6.15 Series Motor

Like the series generator, this machine is a variable-flux machine. Despite this, the back-emf of this motor remains almost constant, from light-load to full-load conditions. This fact is best

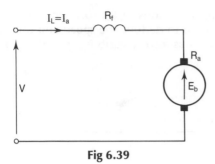

Fig 6.39

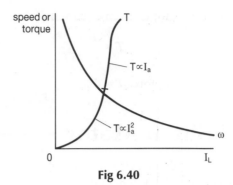

Fig 6.40

illustrated by considering the circuit diagram (Fig. 6.39), the back-emf equation, and some typical values.

$$E_b = V - I_a(R_a + R_f) \text{ volt} \qquad (6.9)$$

Let us assume the following: $V = 200$ V; $R_a = 0.15\ \Omega$; $R_f = 0.03\ \Omega$; $I_a = 5$ A on light-load; $I_a = 50$ A on full-load

Light-load: $E_b = 200 - 5(0.15 + 0.03) = 199.1$ V

Full-load: $E_b = 200 - 50(0.15 + 0.03) = 191$ V

From the above figures, it may be seen that although the armature current has increased tenfold, the back-emf has decreased by only 4%. Hence, E_b remains sensibly constant.

Since $\omega \propto E_b/\Phi$, and E_b is constant, then:

$$\omega \propto \frac{1}{\Phi} \ \ldots \ldots [1]$$

Similarly, $T \propto \Phi I_a$, and since $\Phi \propto I_a$ until the onset of magnetic saturation, then $T \propto I_a^2 \ \ldots \ldots [2]$

and after saturation, $T \propto I_a \ \ldots \ldots [3]$

Using [1] to [3] above, the speed and torque characteristics shown in Fig. 6.40 may be deduced.

Note: From the speed characteristic it is clear that, on very light loads, the motor speed would be excessive. *Theoretically*, the no-load speed would

be infinite! For this reason a series motor must NEVER be started unless it is connected to a mechanical load sufficient to prevent a dangerously high speed. Similarly, a series motor must not be used to operate belt-driven machinery, lifting cranes etc., due to the possibility of the load being suddenly disconnected. If a series motor is allowed to run on a very light load, its speed builds up very quickly. The probable outcome of this is the distintegration of the machine, with the consequent dangers to personnel and plant.

The series motor has a high starting torque due to the 'square-law' response of the torque characteristic. For this reason, it tends to be used mainly for traction purposes. For example, an electric train engine requires a very large starting torque in order to overcome the massive inertia of a stationary train. The power flow diagram appears in Fig. 6.41.

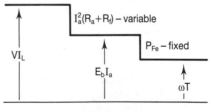

Fig 6.41

6.16 Compound Motors

In order to make maximum use of the supply voltage, the long-shunt configuration is favoured. This is shown in Fig. 6.42. The series winding may, once more, be connected so that its flux either assists or opposes the flux from the shunt winding. The strength of the series field is kept weak compared with the shunt field.

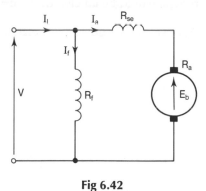

Fig 6.42

▪ Differential compounding ▪

As the load on the machine is increased, the machine draws more current from the supply, and the armature current increases proportionately. Thus the strength of the series field is increased. However, the shunt field is constant, so the overall machine flux is reduced. The result is that the machine will tend to speed up. When the load on a machine is increased, its natural tendency is to slow down. Thus, if the machine is level compounded, this tendency for a drop in speed can be compensated, so that the full-load and no-load speed is the same. The effect of the shunt field is to limit the no-load speed to a safe value. However, the speed will tend to become too high if the machine is overloaded. The differentially compounded motor is used when excellent speed regulation is required. The speed characteristic is shown in Fig. 6.43.

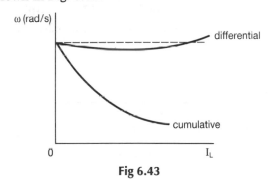

Fig 6.43

▪ Cumulative compounding ▪

Again, the shunt winding is sufficient to limit the no-load speed to a safe value. As the load is increased, so too is the strength of the series field.

Since the fluxes from the two fields are added, the overall machine flux also increases, with a consequent reduction in speed. The machine speed therefore drops considerably as the load is increased. A typical application requiring this type of characteristic would be an electrically driven press.

Figure 6.44 shows the power flow diagram for a compound motor.

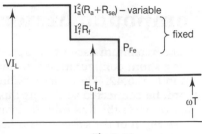

Fig 6.44

6.17 Separately Excited Motor

This machine will have similar characteristics to those of the shunt motor. Its main advantage is the relative ease with which the speed may be varied, by controlling the field current. The direction of

rotation can also be simply changed by reversing the field current. This type of motor is frequently used as a d.c. servo motor, particularly in its split-field form.

■ Worked Example 6.5 ■

Q A d.c. shunt motor develops a torque of 30 Nm at a speed of 850 rev/min when running from a 450 V supply and taking an armature current of 25 A. Calculate its speed and torque when running from a 220 V supply and taking an armature current of 12 A. The armature resistance is 0.8 Ω, and the flux at 220 V is 80% of that at 450 V.

$T_1 = 30$ Nm; $n_1 = 860$ rev/min; $V_1 = 450$ V; $I_{a1} = 25$ A

$\qquad V_2 = 220$ V; $I_{a2} = 12$ A; $\Phi_2 = 0.8\Phi_1$ weber

In general, $n \propto \dfrac{E_b}{\Phi}$; where $E_b = V - I_a R_a$ volt

$\qquad$ so, $E_{b1} = 450 - (25 \times 0.8) = 430$ V

$\qquad$ and, $E_{b2} = 220 - (12 \times 0.8) = 210.4$ V

$$n_1 \propto \frac{E_{b1}}{\Phi_1} \quad \ldots\ldots \text{[1]}$$

$$\text{and } n_2 \propto \frac{E_{b2}}{\Phi_2} \quad \ldots\ldots \text{[2]}$$

and dividing [2] by [1]:

$$\frac{n_2}{n_1} = \frac{E_{b2}\Phi_1}{E_{b1}\Phi_2} = \frac{210.4 \times \Phi_1}{430 \times 0.8 \times \Phi_1} = 0.6116$$

$\qquad$ so, $n_2 = 0.6116 \times n_1 = 0.6116 \times 850$

therefore $n_2 = 520$ rev/min **Ans**

Since $T \propto \Phi I_a$, then $T_1 \propto \Phi_1 I_{a1}$ and $T_2 \propto \Phi_2 I_{a2}$

$$\text{thus, } \frac{T_2}{T_1} = \frac{\Phi_2 I_{a2}}{\Phi_1 I_{a1}} = \frac{0.8 \times \Phi_1 \times 12}{\Phi_1 \times 25}$$

$$T_2 = 0.384 \times 30$$

$\qquad$ so $T_2 = 11.52$ Nm **Ans**

6.18 The Importance of Back-emf

We have seen that the resistance of the armature of a d.c. machine is very small, usually considerably less than one ohm. On the other hand, the supply voltage is often in hundreds of volts. If there was no back-emf to limit the armature current, the latter would be excessive. For example, if $V = 200$ V and $R_a = 0.5$ Ω, then in the absence of a back-emf, $I_a = 400$ A. This value of current would almost certainly cause severe damage to the armature winding. Now, whilst the armature is rotating, a back-emf will be generated, so the

above problem will not exist. However, at the instant of connection to the supply the armature *will* be stationary, and an excessive current would be demanded from the supply. For this reason, some means of limiting the initial starting current must be employed. This effect is achieved by connecting extra resistance in series with the armature. As the machine speed builds up and the back-emf starts to increase, this extra resistance is gradually reduced to zero. The device used for this purpose is the faceplate starter.

6.19 D.C. Faceplate Starter

The general arrangement of the starter, connected between the supply and a shunt motor, is illustrated in Fig. 6.45. The arm is spring-loaded to the 'OFF' position. On the underside of the arm are two plates which make contact with a set of copper studs and a copper strip, when the arm is moved towards the 'ON' position. On the side of the arm is a soft-iron plate which makes contact with the poles of the no-volt relay (NVR) when the arm is in the 'ON' position. Connected between the studs are resistance elements. The supply to the arm contact plates is via the coil of the

overload relay (OLR). The starting procedure is as follows:

The arm is moved so that the supply is connected to the first stud. Thus the armature is supplied via the whole of the starter resistance. The field winding is supplied directly via the copper strip and the coil of the NVR. The armature therefore starts to rotate, and its speed will build up to some steady value. As the arm is moved from stud to stud, the extra resistance in the armature circuit is reduced, and the speed increases. When the last stud is reached, the

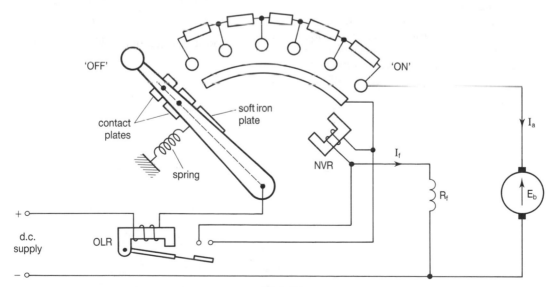

Fig 6.45

armature is connected directly to the supply, and the machine will be running at its normal no-load speed. With the arm in the 'ON' position, the flux produced by the coil of the NVR will be sufficient to hold the arm in this position, by attracting the soft-iron plate.

The NVR, the OLR and the spring on the arm provide protection for the motor, as follows. If the supply should fail whilst the machine is running normally, the coil of the NVR will de-energise, thus releasing the plate, and the spring returns the arm to the 'OFF' position. This procedure is necessary, otherwise, if the arm remained in the 'ON' position, when the supply is restored the stationary armature would be connected directly to the supply. The NVR therefore ensures that the correct starting procedure must be repeated after failure of the supply. The OLR has a hinged armature which carries an electrical contact. Whilst the line current from the supply is equal to or less than 1.5 times the normal full-load value, the flux produced by the OLR coil is insufficient to pull-in its armature. If the machine is seriously overloaded then the OLR armature will close, thus short-circuiting the NVR coil. The result is that the NVR de-energises and the spring returns the arm to the 'OFF' position.

6.20 Efficiency of D.C. Machines

As with any machine, the efficiency may be calculated from the ratio of the output power to the input power. In addition, the losses involved can be split between those that are unvarying (fixed losses) and those that will vary according to the loading placed on the machine. The condition for maximum efficiency is again achieved when the variable losses are equal to the fixed losses. The power flow diagram for a given type of machine identifies this division of losses. These diagrams can prove very useful when solving machine problems, and it is recommended that they be referred to when calculations are undertaken.

Provided that the armature and field circuit resistances are known, then the copper losses, P_{Cu}, can be easily calculated for given load conditions. The iron, and friction losses, P_{Fe}, are determined by conducting a simple test, as follows:

The machine is run light (on no-load) as a motor, being supplied with its normally rated terminal voltage. Under these conditions the armature current will be very small compared to its normal full-load value. For this reason the variable losses are negligible, and the input power to the machine may be taken as the machine's fixed losses. For the different types of machine, normally used as either motors or generators, the above simple test yields the following:

Shunt machine:	$VI_L = P_{Fe} + I_f^2 R_f$ watt
Series machine:	$VI_L = P_{Fe}$ watt
Compound machine:	$VI_L = P_{Fe} + I_f^2 R_f$ watt

■ **Worked Example 6.6** ■

  A 240 V shunt motor, running on no-load and at normal speed, takes an armature current of 2.4 A from a 240 V supply. The resistance of the field circuit is 240 Ω, and that of the armature is 0.25 Ω. Calculate the motor output power and efficiency when drawing a line current of 35 A from the supply.

A

From the data given regarding the no-load conditions, the machine's fixed losses may be determined as follows. The circuit for these conditions is shown in Fig. 6.46.

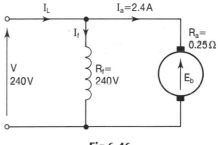

$$\text{On no-load: } I_f = \frac{V}{R_f} \text{ amp} = \frac{240}{240} = 1 \text{ A}$$

$$I_L = I_a + I_f \text{ amp} = 2.4 + 1 = 3.4 \text{ A}$$

$$\text{fixed losses} = VI_L \text{ watt} = 240 \times 3.4$$

$$\text{so fixed losses} = 816 \text{ W}$$

Fig 6.46

The circuit conditions on load are shown in Fig. 6.47.

$$\text{On load: } I_a = I_L - I_f \text{ amp} = 35 - 1 = 34 \text{ A}$$

$$P_{Cu} = I_a^2 R_a \text{ watt} = 34^2 \times 0.25 = 289 \text{ W}$$

$$\text{total losses} = 816 + 289 = 1105 \text{ W}$$

$$\text{input power, } P_i = VI_L \text{ watt} = 240 \times 35 = 8400 \text{ W}$$

$$\text{output, } P_o = P_i - \text{losses} = 8400 - 1105$$

$$\text{therefore, } P_o = 7.295 \text{ kW Ans}$$

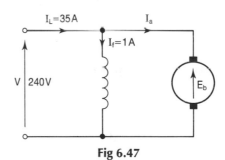

$$\eta = \frac{P_o}{P_i} \times 100\% = \frac{7295}{8400} \times 100\%$$

$$\text{so } \eta = 86.85\% \text{ Ans}$$

Fig 6.47

■ Worked Example 6.7 ■

 A shunt generator produces a full-load output of 10 kW at a terminal voltage of 250 V. The resistances of the armature and field circuits are 0.5 Ω and 125 Ω respectively. After running the machine as a motor on no-load, it was found that the iron and friction losses are 500 W. Determine (a) the power required at the driving shaft to provide the full-load output, (b) the input torque at full load if the driving speed is 900 rev/min, and (c) the full-load efficiency.

A ───────────────────────────

The circuit diagram is shown in Fig. 6.48.

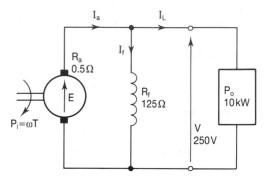

Fig 6.48

(a)

$$I_L = \frac{P_o}{V} \text{ amp} = \frac{10^3}{250} = 40 \text{ A}$$

$$I_f = \frac{V}{R_f} \text{ amp} = \frac{250}{125} = 2 \text{ A}$$

$$I_a = I_L + I_f \text{ amp} = 42 \text{ A*}$$

$$E = V + I_a R_a \text{ volt} = 250 + (42 \times 0.5) = 271 \text{ V}$$

generated power,

$$EI_a = 271 \times 42 = 11382 \text{ W}$$

$$P_i = EI_a + P_{Fe} \text{ watt} = 11382 + 500$$

so $P_i = 11.882$ kW **Ans**

(b)

$$\omega = 2\pi n \text{ rad/s; where } n = 900/60 \text{ rev/s}$$

$$= 2 \times \pi \times \frac{900}{60} = 94.25 \text{ rad/s}$$

$$\omega T = P_i = 11882 \text{ W}$$

so $T = \dfrac{11882}{94.25} = 126.1$ Nm **Ans**

(c)

$$\eta = \frac{P_o}{P_i} \times 100\% = \frac{10000}{11882} \times 100$$

hence $\eta = 84.16\%$ **Ans**

*alternatively, $P_{Cu} = I_a^2 R_a + I_f^2 R_f$ watt

$$= (42^2 \times 0.5) + (2^2 \times 125)$$

so $P_{Cu} = 1382$ W

total losses $= P_{Cu} + P_{Fe}$ watt $= 1382 + 500 = 1882$ W

$$P_i = P_o + \text{losses} = 11.882 \text{ kW } \textbf{Ans}$$

6.21 Simple Methods of Speed Control

The speed of a d.c. motor may be controlled by varying either the armature current or the field current. With self-excited machines, the only way these two currents can be varied is to reduce them from their full-load values. This is normally achieved by some form of variable resistance, connected in the relevant circuit.

■ Field regulator ■

This method is suitable for use only with shunt and compound machines. The regular consists of a variable resistor connected in series with the shunt field winding. When this resistance is increased, the current decreases, thus weakening the flux. This results in an increase of speed. Using this method it is possible to increase the speed to three or four times that at full excitation.

■ Controller ■

This is a variable resistor connected in series with the armature, similar to starter resistance. As this resistance is increased, so the back-emf is reduced, resulting in a decrease of speed. Speeds from zero to full speed are possible, but this method is very wasteful of power, and hence severely reduces the efficiency of the motor.

■ Separate excitation ■

As stated in Section 6.17, the speed and direction of rotation of a separately excited motor is relatively easy to control, simply by varying the field current.

The speed of a series (or universal) motor used in small domestic appliances, such as a food mixer, is often achieved by switching different field coils into and out of circuit. This results in a predetermined speed for each field coil.

6.22 Stepper Motors

Stepper or stepping motors cannot readily be classified as either d.c. or a.c. machines. Logically they should be classified as digital machines, since they are controlled by electronic circuitry that in turn controls the sequence of pulses fed to the motor coils. The other main feature of this type of motor is that the rotation occurs in discrete angular steps. The step angle may be either large (say 30°) or small, depending upon the application. A step angle of 1.8° is quite common, which allows 200 steps per revolution. The stepping speed is controlled by the frequency or pulse rate of the control signal, and up to about 800 steps/second can be achieved. Since the angular position of the machine's rotor can be controlled with great precision, stepper motors are used in applications such as robotics, numerically controlled machines, head positioners for floppy and hard disc drives, X-Y plotters, video cassette recorder heads etc.

A simple form of stepper motor is illustrated in Fig. 6.49. The rotor is made from a 'soft' magnetic material, and carries no electrical windings. The number of rotor teeth will be different to the number of stator poles. In the case shown, the stator poles are displaced by 45°, whereas the rotor teeth are displaced by 60°. As will be shown, this will result in a step angle of 15°. Windings are mounted on the stator poles in such a way that they can produce magnetic flux in the axes AA′, BB′, CC′, etc. when they are energised from the control circuit.

Consider that stator coils AA′ have been energised, and the rotor has aligned with the resulting flux, as shown in the diagram. If AA′ is de-energised as BB′ is energised, the flux axis rotates through 45° clockwise. Rotor teeth 2 and 5 will now align with this flux, thus turning the rotor

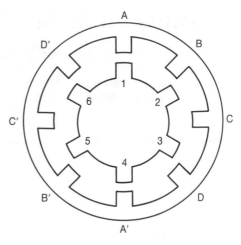

Fig 6.49

through a 15° step anticlockwise. If BB′ is then de-energised as CC′ is energised, rotor teeth 3 and 6 will align with CC′; and so on. The direction of rotation is simply reversed by reversing the sequence in which the stator coils are energised.

Assignment Questions

1 A four-pole wave wound machine has 400 armature conductors and is driven at 900 rev/min. If the generated emf is 200 V, determine the useful flux per pole.

2 A six-pole machine has 420 armature conductors. If the flux per pole is 35 mWb, determine the speed of rotation required in order to generate an emf of 250 V when the armature is (a) lap-wound and (b) wave-wound.

3 A shunt generator supplies a current of 85 A at a terminal p.d. of 380 V. Calculate the generated emf if the armature and field resistances are 0.4 Ω and 95 Ω respectively.

4 A generator produces an armature current of 50 A when generating an emf of 400 V. If the terminal p.d. is 390 V, calculate (a) the value of the armature resistance, and (b) the power loss in the armature circuit.

5 A d.c. shunt generator supplies a 50 kW load at a terminal voltage of 250 V. The armature and field circuit resistances are 0.15 Ω and 50 Ω respectively. Calculate (a) the generated emf, and (b) the efficiency if the mechanical and iron losses total 2.45 kW.

6 A short-shunt compound generator has armature, shunt field and series field resistances of 0.8 Ω, 50 Ω and 0.6 Ω respectively. The shunt field copper loss is 312.5 W; the series field copper loss is 240 W, and the terminal voltage is 250 V. Determine (a) the power output, and (b) the generated emf.

7 A shunt-wound machine has armature and field circuit resistances of 0.05 Ω and 105 Ω respectively. When the machine is driven as a generator at 650 rev/min it produces an output of 50 kW at a terminal voltage of 420 V. Determine the speed at which it would run as a motor when taking an input of 50 kW from a 420 V d.c. supply.

8 A series motor runs at 600 rev/min when connected to a 240 V d.c. supply, and drawing 30 A. Calculate its speed when drawing a current of 42 A from a 300 V supply. The combined armature and field resistance is 0.48 Ω. You may assume that the flux remains directly proportional to the field current.

9 A four-pole motor has its armature lap-wound with 1000 conductors, and it runs at 1000 rev/min when taking an armature current of 40 A from a 250 V supply. Given that the armature resistance is 0.18 Ω, calculate (a) the useful flux per pole, and (b) the gross torque developed by the armature.

10 A shunt generator has a full-load output of 10 kW at a terminal voltage of 240 V. The armature and field resistances are 0.5 Ω and 150 Ω respectively. The mechanical and iron losses total 500 W. Determine (a) the input driving power required at full-load, (b) the full-load efficiency, and (c) the armature current that results in maximum efficiency.

11 A d.c. motor was tested by connecting its shaft to the brake pulley shown in Fig. 6.50. The suspended mass was 25 kg and the reading on the spring balance was 60 N when the motor was running at 950 rev/min. Given that the diameter of the brake pulley was 350 mm, and the motor input was 20 A at a terminal voltage of 200 V, calculate (a) the motor output, and (b) its efficiency at this load.

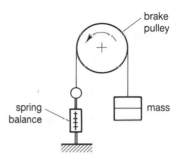

Fig 6.50

12 A d.c. shunt generator has a full-load output of 180 kW at 250 V. The machine was tested as detailed below, yielding the results shown.

(i) When running light as a motor at full speed, it drew a line current of 35 A from a 250 V supply, and the field current was 10 A.

(ii) With the machine stationary, an emf of 3 V across the armature produced a current of 200 A.

Using these results, determine the full-load efficiency.

13 The current drawn by a 450 V shunt motor, when running light, is 8 A. The armature and field resistances are 0.15 Ω and 250 Ω respectively. Calculate (a) the input power and the efficiency when a current of 120 A is drawn from this 450 V supply, and (b) the armature current at which maximum efficiency occurs.

Suggested Practical Assignments

Assignment 1

To obtain the open-circuit and output characteristics for a separately excited d.c. generator

Apparatus:
1 × separately excited generator (fhp machine)
1 × d.c. supply (e.g. 110 V)
2 × single-pole switch

2 × rheostats
2 × ammeter
1 × voltmeter
1 × constant-speed drive motor

Method:

1 Connect the circuit shown in Fig. 6.51, with both switches left open and both rheostats set to maximum resistance.

2 Operate the drive motor, measure and tabulate any generated emf.

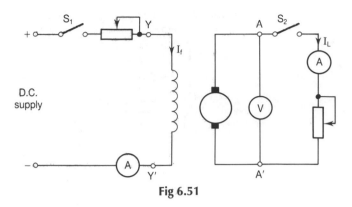

Fig 6.51

3 Close switch S_1, measure and tabulate the values of field current and generated emf.

4 Increase the field current, in discrete steps, and tabulate these current settings together with the corresponding emf generated. Continue this procedure until the generator is producing its normally rated emf.

5 Now close S_2, measure and tabulate the corresponding values of load current and terminal voltage.

6 Increase the load current, in discrete steps, tabulating the current and terminal voltage at each step. Continue this procedure up to the rated full-load output for the machine.

7 From your tabulated data, plot a graph of generated emf versus field current, and a graph of terminal voltage versus load current.

8 Complete an assignment report, including comments regarding the shapes of the two characteristics compared with those predicted by theory.

Assignment 2

To obtain the output characteristics for a shunt-wound d.c. generator.

Apparatus:
1 × shunt generator
1 × drive motor
1 × rheostat
1 × ammeter

1 × voltmeter
1 × single-pole switch

Method:

1 Connect the circuit as in Fig. 6.52, leaving the

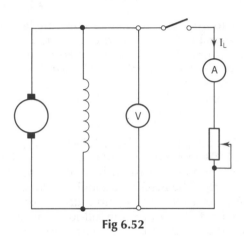

Fig 6.52

switch in the open position, and the rheostat set to its maximum resistance setting.

2 Drive the generator and note the value of generated voltage. Note that this voltage should build up rapidly to its normally rated output value. If self-excitation does not occur, switch off the drive motor, reverse the connections to the generator shunt field, and restart the motor. If the generator still fails to self-excite, consult your lecturer.

3 Close the switch, note and tabulate the corresponding values of load current and terminal voltage.

4 Increase the load current, in steps, noting the values of load current and terminal voltage at each step. Continue this procedure up to the full-load output of the machine.

5 If possible, increase the load current beyond the full-load value, and observe the affects on both this current and the terminal voltage.

6 Prepare an assignment report. This should include a plotted graph of the generator output

characteristic, and comments regarding its shape etc.

Assignment 3

To obtain the output characteristics for a compound generator, using both cumulative and differential compounding.

Apparatus:
1 × compound generator

Remainder of apparatus as for Assignment 2.

Method:
As for Assignment 2, except that the procedures will be carried out for both cumulative and differential compounding. In order to change from one form of compounding to the other, simply reverse the connections to the series field winding. The circuit is shown in Fig. 6.53.

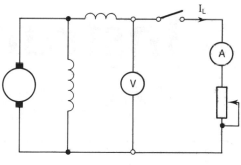

Fig 6.53

▪ 7 Measuring Instruments and Measurements ▪

This chapter is intended to help you understand the principal features and limitations of some commonly used measuring instruments. On completion you should be able to:

1] Appreciate the sources of error that can occur when taking measurements.

2] Predict the loading effects and frequency response characteristics of instruments.

3] Predict the affect of distorted waveforms on detectors used in electronic instruments.

4] Calculate the power consumed by instruments connected into circuits.

5] Appreciate the basic principles of a.c. bridge measurements.

6] Select and correctly use appropriate instruments for a variety of measurements.

7.1 Introduction

The operating principles of moving coil and rectifier moving coil multimeters; Wheatstone bridge; slidewire potentiometer; and CRO were covered in Volume 1 of this title. In addition, you will by now, have gained some practical experience of simple voltage, current and resistance measurements, using the above instruments. As part of this experience you may well have come to the important realisation that *no* measurement should ever be considered as being absolutely correct. Any measurement that is made will involve a number of possible sources of error.

▪ Calibration errors ▪

The calibration of an instrument is the process whereby its indicated values are compared against a known standard. Since no instrument can be made perfect in all respects, there will be discrepancies between the indicated values and the 'true' values. These discrepancies are known as the

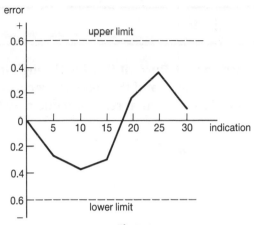

Fig 7.1

calibration errors, and may be shown in the form of a graph, as in Fig. 7.1. There will be some allowable limits of error specified for the instrument. The range of these limits determines the 'grade' which the instrument is allocated; the lower the limits of error the 'higher' the grade. Figure 7.1 could represent the calibration graph

for a 0 to 30 V voltmeter, with a maximum allowable error of ±0.6 V. The accuracy of an instrument is always quoted in terms of the maximum allowable error, usually as a percentage of the full-scale deflection (fsd). Thus for the above example, the accuracy of the instrument would be quoted as being ±2% of fsd. It is worth noting that if the instrument was in error by ±0.6 V at an indicated value of, say, 10 V, then the actual percentage error at this indication would be:

$$\pm \frac{0.6}{10} \times 100\% = \pm 6\%$$

This is because the allowable error of 0.6 V is a much more significant proportion of 10 V than it is of 30 V. For this reason, when using pointer-on-scale instruments, it is advisable to use an instrument (or the appropriate scale on the instrument) so that the indicated value is as near to fsd as is possible. When this is done, the calibration error is minimised. The affect of this form of error is illustrated in Example 7.1.

■ Worked Example 7.1 ■

Q The value of a 50 Ω resistor is to be verified by measuring the current flowing through it and the corresponding p.d. across it. Details of the two meters, and the readings recorded are shown below. Determine the range of values between which the actual resistor value may be said to lie.

Voltmeter: accuracy ±2% of fsd; range 0–100 V; indicated reading 75 V

Ammeter: accuracy ±2% of fsd; range 0–3 A; indicated reading 1.5 A.

A

Possible voltmeter error = ±0.02 × 100 = ±2 V
therefore possible p.d. = 75 ± 2 = 73 V to 77 V

possible ammeter error = ±0.02 × 3 = ±0.06 A
therefore possible current = 1.5 ± 0.06 = 1.44 A to 1.56 A

Since $R = V/I$ ohm, then the maximum error of measurement will occur when one meter error is at its upper limit, and that for the other meter is at its lower limit.

i.e. when $V = 77$ V and $I = 1.44$ A
or when $V = 73$ V and $I = 1.56$ A

hence, $R = \dfrac{77}{1.44} = 53.472 \ \Omega$;

or $R = \dfrac{73}{1.56} = 46.795 \ \Omega$

Therefore, the resistor value may be said to have a value in the range 46.795 Ω and 53.472 Ω **Ans**

■ Systematic errors ■

This form of error is either avoidable, or may be taken into account when a measurement is made. Failure to ensure that an instrument is correctly 'zeroed' before connecting it to a circuit is a common but easily avoidable systematic error. The loading effect of an instrument, and the points in the circuit to which it is connected may also be taken into account in order to minimise this form of error. This form of error is illustrated in Example 7.2.

▪ Worked Example 7.2 ▪

Q A 5 kΩ resistor is connected across a 10 V supply. The p.d. across the resistor and the current flowing through it are measured as shown in Fig. 7.2. The voltmeter has a total resistance of 20 kΩ, and that of the ammeter is 5 Ω. Assuming negligible calibration errors, let us determine the systematic errors in the two meter readings, attributable to their manner of connection.

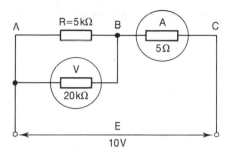

Fig 7.2

A

With neither meter connected, then the 'true' values for p.d. and current will be 10 V and 2 mA respectively. After connection:

$$R_{AB} = \frac{5 \times 20}{2 + 20} \text{ k}\Omega = 4 \text{ k}\Omega$$

$$R_{AC} = 4.005 \text{ k}\Omega; \text{ and}$$

$$V = \frac{R_{AB}}{R_{AC}} \times E = \frac{4}{4.005} \times 10 = 9.888 \text{ V}$$

This would be the p.d. across the voltmeter, and assuming that it could be accurately read down to three decimal places, then the error would be:

$$\frac{9.888 - 10}{10} \times 100\% = -1.12\%$$

This degree of error would be acceptable. Indeed, the voltmeter reading would most likely be interpreted as 10 V.

The circuit current, $I = \frac{E}{R_{AC}}$ amp

$$= \frac{10}{4005} = 2.497 \text{ mA}$$

This is the value of current measured by the ammeter, so the error in this reading would be:

$$\frac{2.497 - 2}{2} \times 100\% = +25\%$$

This degree of error is clearly unacceptable. However, if

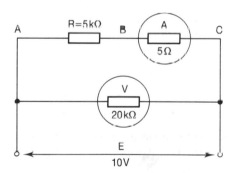

Fig 7.3

the voltmeter was connected between A and C, as in Fig. 7.3, then:

Voltmeter reading = 10 V, i.e. zero systematic error

Current through ammeter = $\frac{E}{R_{ABC}}$ amp = $\frac{10}{5005}$ = 1.998 mA

so ammeter error = $\frac{1.998 - 2}{2} \times 100\% = -0.1\%$

This degree of error is most acceptable, so this alternative connection of the voltmeter is obviously better at reducing the total systematic error. It is left to the reader to verify that if the voltmeter resistance was (say) 10 MΩ, then the systematic error introduced by the voltmeter connection of Fig. 7.2 would be negligible. This clearly illustrates one advantage of a digital voltmeter compared with a moving coil instrument. That is, a perfect voltmeter would have an infinite resistance (draw zero current). On the other hand, a perfect ammeter would have zero resistance, and hence cause no additional potential drop in the circuit.

■ Observational and random errors ■

All other forms of measurement error may be included under this heading. Observational (human) error can occur in a number of ways. When using a pointer-on-scale instrument, the pointer often comes to rest between two marked graduations on the scale. The reading recorded is therefore open to interpretation by the individual observer, and different observers will tend to give slightly different interpretations. Occasionally there may be very large observational errors. This is fairly common when a student first uses a multimeter such as an AVO, and inadvertently reads the indication from the wrong scale, or misinterprets the scale setting applied by the rotary range switch. This problem with the range selection can also occur with digital instruments. With this form of instrument there can be no 'argument' as to the digits displayed, but the displayed value is normally subject to a ±1 digit random error. Temperature and other environmental factors may also be the cause of random errors in measurements.

7.2 Accuracy and Sensitivity

These two measurement terms are often confused with each other, despite the fact that they describe different features of a measuring instrument or system.

■ Accuracy ■

This has already been described as the error referred to the fsd. Another way to define accuracy would be to say that it represents the closeness to the true value that is obtainable.

■ Sensitivity ■

The sensitivity, discrimination or resolution of an instrument describes the smallest change of the measured quantity (measurand) that can be discerned on the display. Thus sensitivity may be defined as:

Comparing accuracy and sensitivity, it can be said that a sensitive instrument is not necessarily accurate; but an accurate instrument needs to be sensitive.

$$\frac{\text{change of indication (output)}}{\text{change of measured quantity (input)}}$$

7.3 Total Measurement Error

The in-depth study of the theory of errors is beyond the scope of the course of study being undertaken here. However, a simplified coverage of the most common techniques will be outlined, in order to give an appreciation of the result of total error.

▪ Addition and subtraction of measurements ▪

Consider two meter readings which are to be added or subtracted to obtain a final result. Let D_1 and D_2 be the two displayed readings, and e_1 and e_2 the respective maximum errors. The sum of the two readings, D_s, and the total error is:

$$D_s = D_1 + D_2 \pm (e_1 + e_2)$$

and the difference,

$$D_d = D_1 - D_2 \pm (e_1 + e_2)$$

Hence the total error in the sum or difference of two or more readings is equal to the sum of the individual errors. Note that e_1 and e_2 are the absolute errors in the readings, and not the quoted accuracies of the instruments. For example,

suppose that two p.d.s are measured in a series d.c. circuit, and the sum of the two represents the applied voltage. The voltmeter used has an accuracy of $\pm 2\%$ and an fsd of 100 V. If the two p.d.s indicated are $V_1 = 60$ V and $V_2 = 20$ V, the applied voltage and associated total error is obtained thus:

Since meter has an accuracy of $\pm 2\%$ of fsd, then absolute possible error is ± 2 V. The possible values for V_1 and V_2 are,

$$V_1 = 60 \pm 2 \text{ V and } V_2 = 20 \pm 2 \text{ V}$$
$$\text{so, } V = V_1 + V_2 = (60 + 20) \pm (2 + 2) \text{ V}$$
$$= 80 \pm 4 \text{ V}$$

Thus the total possible error is ± 4 V or $\pm 5\%$.

▪ Multiplication and division of measurements ▪

Consider two measurements and their associated errors as follows:

$$D_1 \pm e_1 = D_1 \left(1 \pm \frac{e_1}{D_1}\right) = D_1 (1 \pm \delta_1)$$

and similarly, $D_2 (1 \pm \delta_2)$

where $\delta_1 = e_1/D_1$; $\delta_2 = e_2/D_2$ are the *fractional* errors. The product of the two readings yields the result,

$$D_p = D_1 D_2 (1 \pm \delta_1)(1 \pm \delta_2)$$
$$= D_1 D_2 [1 \pm (\delta_1 + \delta_2) + \delta_1 \delta_2]$$

but since δ is generally a very small quantity, then the sum $(\delta_1 + \delta_2)$ is normally very much greater than the product $\delta_1 \delta_2$.

Therefore, $\qquad D_p = D_1 D_2 [1 \pm (\delta_1 + \delta_2)]$

So, for a product of two measurements, the total fractional error is the sum of the individual fractional errors. Similarly, it can be shown that when dividing one reading by another, the quotient,

$$D_q = \frac{D_1}{D_2}[1 \pm (\delta_1 + \delta_2)]$$

■ Power or root of a measurement ■

Since a root is simply a fractional power, then the same technique applies to both. For example, the square-root of a quantity expressed as a power is,

$$\sqrt{y} = y^{1/2}$$

It is found that the total fractional error in raising to a power n equals n times the original fractional error, such that:

reading raised to the power $n = D^n(1 \pm n\delta)$

7.4 Wattmeter Corrections

When a wattmeter is used to measure the power in a circuit, systematic errors are introduced. The amount of error thus introduced depends upon the way in which the wattmeter is connected into the circuit. Consider Fig. 7.4, which shows a

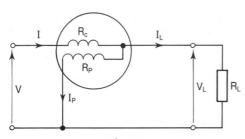

Fig 7.4

wattmeter connected between a supply and its load. The wattmeter reading depends on the p.d. across its voltage coil, and the current flowing through its current coil. However, the current through the current coil is the load current, I_L, *plus* the small current I_p drawn by the voltage or pressure coil. The wattmeter reading is therefore:

$$P = V_L(I_L + I_p) \text{ watt}$$

$$= V_L I_L + V_L I_p$$

$$= \text{load power} + V_L I_p \text{ watt.}$$

The wattmeter reading is therefore too high, by the amount of power dissipated in the pressure coil. This coil of the wattmeter will have a relatively high resistance, so I_p will be correspondingly small. Thus, if $I_L \geqslant 10 \times I_p$, then the error introduced is negligible. If, on the other hand, the load current is of a comparable value to I_p, then the error can be significant. In this situation the wattmeter reading should be

corrected by subtracting the pressure coil power from the meter reading. This is possible provided that the resistance of the pressure coil is known, such that:

$$P_p = \frac{V_L^2}{R_p} \text{ watt, or } P_p = I_p^2 R_p \text{ watt}$$

The wattmeter may be connected so that the pressure coil is connected to the supply side of the circuit as in Fig. 7.5. In this situation the current

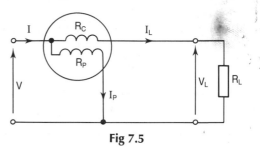

Fig 7.5

through the current coil will be the true load current. However, the p.d. across the pressure coil is the load p.d. plus the potential drop due to the resistance of the current coil. The total wattmeter reading is therefore:

$$P = (V_L + IR_c)I_L \text{ watt}$$

$$= V_L I_L + I_L^2 R_c$$

$$= \text{load power} + I_L^2 R_c$$

This means that the meter reading will again be high, and may be corrected provided that the current coil resistance is known. This form of connection will minimise the error when the load current (and hence the current coil p.d.) is relatively small.

7.5 Electronic Voltmeter

It has been shown that the main source of error for analogue voltmeters, such as the AVOmeter, occurs when measuring the p.d. across a high value resistor. This error is due to the loading effect of the meter's input resistance. Additionally, when used on the a.c. ranges the upper frequency limit will be in the order of 20 kHz. An electronic voltmeter reduces these sources of error by employing a transistor amplifier circuit. This amplifier is normally based on a field effect transistor (FET), which can have an input resistance in the order of hundreds of megohms. This amplifier may also be connected as a DC (directly coupled) amplifier, which enables it to have a bandwidth extending from 0 Hz up to many megahertz. Since this amplifier can have considerable voltage gain, there is a limitation on the size of input voltage that may be applied to it. For this reason a switchable attenuator network is interposed between the meter terminals and the amplifier input terminals. One disadvantage of this attenuator network is to reduce the overall input resistance of the instrument to the region of 10–20 MΩ. However, this is still far greater than that of an AVO, so the loading effect of the electronic voltmeter is minimal. Another advantage of employing an amplifier is that an electronic voltmeter may be capable of measuring voltages as low as microvolts. The amplified voltage is normally displayed on a conventional moving coil movement. For this reason, when the instrument is used for a.c. measurements, a rectifier is switched into the input circuit. A much simplified diagram of such an instrument is shown in Figs. 7.6(a) and (b).

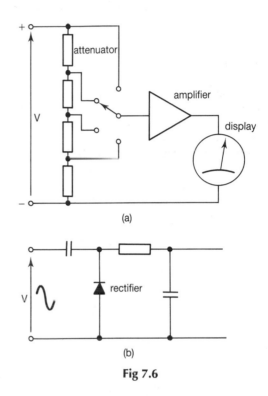

(a)

(b)

Fig 7.6

7.5 Digital Voltmeter (DVM)

As with the electronic voltmeter, a DVM has a high input resistance because it employs a transistor amplifier and input attenuator. The amplifier is usually in the form of an operational amplifier connected as an integrator. This, coupled with an oscillator, a counter and associated logic control circuitry forms an analogue-to-digital converter (ADC) circuit. Due to the presence of frequency dependent components such as capacitors, the bandwidth of such an arrangement is much less than that achievable from an electronic voltmeter. Indeed, some of the commonly used (and 'cheaper') DVMs may have a bandwidth of only 2–4 kHz. It is therefore most important to consult the instrument data sheet to check on this aspect. Since an AVO has a better frequency response than the above figure, then this instrument could prove to be more reliable for a.c. measurements at frequencies greater than 2 kHz. For power frequency and d.c. measurements any DVM will normally be preferable to an analogue instrument.

7.6 Complex Waveforms

The concept of complex waveforms is dealt with in Volume 1, Chapter 9. The implications regarding the bandwidth required for the transmission of complex waveforms was also described. The measurement of a complex waveform voltage using an analogue moving coil meter will normally result in a large error. This is because such an instrument is calibrated to indicate the rms value of a sinusoid, by employing a correction factor equal to the form factor for a sinewave. Depending upon the type of circuitry employed, electronic and digital instruments may introduce either zero error or a large error due to the complexity of the waveform. An instrument that is quoted as being a 'true rms' meter should introduce no error due to waveshape. However, in the case of this group of instruments, the bandwidth of the internal amplifier may introduce errors.

The best means of measuring a complex waveform is probably by the use of an oscilloscope. This gives the added advantage that the actual waveshape can also be observed. However, for the displayed waveform to be a faithful reproduction of the measured waveform *all* of the harmonics present in the original must be amplified (or attenuated) by the same amount. If this is not the case then the waveshape will be changed. Thus the amplifiers employed must have a very wide bandwidth, with a flat frequency response curve. The other main problem concerns any time delay (phase shift) that the circuitry imposes on different harmonics. A slight change of phase of a harmonic can result in a dramatic change in waveshape. For these reasons the amplifiers used in oscilloscopes are usually directly coupled. This avoids the use of coupling capacitors and resistors (which can introduce time delays and phase shift) and also allows the amplification of signals down to 0 Hz. Since most complex waveforms contain a d.c. component, it is important that this is treated in the same way as all the harmonics. In order to obtain an appreciation of the changes produced in a complex wave due to variations in the amplitude and phase of harmonics, it is suggested that experimentation with a waveform synthesiser and oscilloscope be undertaken.

7.7 Measurement of Phase and Frequency

These two characteristics of alternating quantities may be measured by meters designed specifically for this, i.e. a phasemeter and a frequency meter/counter. These instruments are not always readily available, and these measurements are more often performed with an oscilloscope, as will now be described.

The measurement of the phase relationship between two waveforms (of the same frequency) may be carried out with a CRO using either one of two methods.

▪ Dual-trace method ▪

The two waveforms are displayed on the screen, and by means of the 'Y'-shift controls the two traces are aligned along a common horizontal axis, as shown in Fig. 7.7. The time intervals T and t are then measured, using the graticule and timebase setting. Since the periodic time T of a waveform corresponds to 360°, then the phase angle ϕ may be determined from:

$$\frac{\phi}{360} = \frac{t}{T}$$

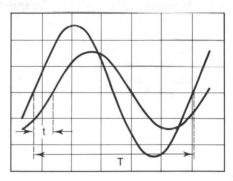

Fig 7.7

▪ Lissajous figures ▪

In this method the CRO timebase is turned off, and one of the two waveforms is connected to the 'X' input. The second waveform is connected to the 'Y'-amplifier input as normal. Depending upon the phase angle between the two waveforms the resulting display will be either a straight line, a circle, or an ellipse. Examples are shown in Fig. 7.8, where the phase relationships are:

(a) $\phi = 0°$ (in phase)
(b) $\phi = 90°$
(c) $\phi = 180°$ (in antiphase)
(d) $0° < \phi < 90°$

Where the display is an ellipse, the phase angle may be calculated by measuring the dimensions A and B, as in Fig. 7.9, and applying the following:

$$\sin \phi = \frac{A}{B}$$

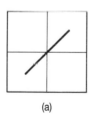

(a)

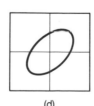

(b)

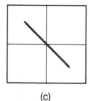

(c)

(d)

Fig 7.8

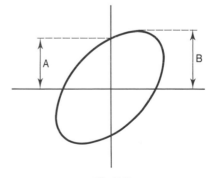

Fig 7.9

7.10 A.C. Bridges

An a.c. bridge utilises the same basic principles as those of the Wheatstone bridge used for the d.c. measurement of resistance. The bridge supply will of course be an alternating supply, and the detector in the central limb must be sensitive to a.c. The balance condition occurs when the current through the detector is zero. The balance condition may also be expressed in terms of the ratio of the impedances of the four outer limbs, as below. The basic arrangement is shown in Fig. 7.10.

$$\frac{Z_1}{Z_2} = \frac{Z_3}{Z_4}$$

One of the limbs will contain the impedance to be measured (e.g. an inductor), and the other arms will contain variable reactance and resistance elements. Obtaining the balance condition is more difficult than for the simple Wheatstone bridge because the potentials at B and D must be equal in both amplitude and phase. There are a number of different forms of a.c. bridge for the measurement of inductors and capacitors, including the Owen, Hay's, Maxwell, and Schering bridges. Each of these may be used to measure either inductance or

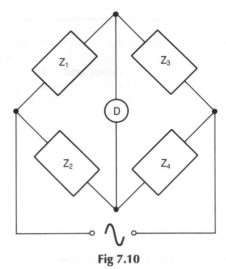

Fig 7.10

capacitance. More commonly, 'Universal' bridges that may be used to measure inductance, capacitance, resistance and Q-factor are employed. It is suggested that the student carries out measurements using a Universal bridge, since the practical experience thus gained will be far more instructive than a description of the procedure given here.

▪ 8 Control Principles ▪

This chapter builds upon and extends the basic principles of control systems that were dealt with in Volume 1. On completion you should be able to:

1] Appreciate the different forms of system element, and analogous elements.

2] Identify a range of first-order systems, and describe their response to step disturbance inputs.

3] Distinguish between first-order and second-order systems.

4] Describe the response of second-order systems to both step inputs and forcing functions, particularly with reference to system damping.

5] Explain the effect of mechanical resonance, and methods of reducing its affects.

6] Understand the concepts of proportional, differential and integral control strategies.

8.1 System Elements and Analogies

All types of system consist of elements or components that are interconnected in such a way that the 'behaviour' of each element has an affect on the 'behaviour' of the whole system. Some system elements are capable of storing energy whilst others can only dissipate energy. Of the energy storage elements, some are associated with kinetic energy (or its equivalent); the others being associated with potential energy (or its equivalent). In order to put these concepts into perspective consider Table 8.1, which shows the basic mechanical and electrical elements classified into the three types mentioned above.

Table 8.1 Energy storage and dissipative elements

System	'KE' storage	'PE' storage	Dissipative
Mechanical	mass $\left(\dfrac{1}{2}mv^2\right)$	spring $\left(\dfrac{1}{2}ks^2\right)$	damper $\left(R\dfrac{\mathrm{d}s}{\mathrm{d}t}\right)$
Electrical	inductor $\left(\dfrac{1}{2}LI^2\right)$	capacitor $\left(\dfrac{1}{2}CV^2\right)$	resistor $\left(R\dfrac{\mathrm{d}q}{\mathrm{d}t}\right)$

Considering Table 8.1, the following analogies can be made:

Both the mass and the inductor store energy in a form associated with movement. In the case of the mass, this will be kinetic energy by virtue of its velocity. For the inductor, the energy is stored in the magnetic field produced by the current flowing. Remember, $v = ds/dt$ metre per second; and $I = dq/dt$ coulomb/second (amp), i.e. the current is the rate at which charge is displaced.

Both the spring and the capacitor store energy by virtue of position or potential—that is, potential energy. For example, the further the spring is displaced the more energy it will store. Similarly, the more charge placed on the plates of a capacitor the larger its p.d., hence more energy stored.

Both the damper and the resistor have a 'slowing down' effect in their respective systems. As a result they cause heat energy to be dissipated.

The analogies can now logically be extended to the system variables and constants such that:

voltage, $V \equiv$ force, F
charge, $Q \equiv$ displacement, s
current, $I \equiv$ velocity, v
resistance, $R \equiv$ damper coefficient, R
inductance, $L \equiv$ mass, m
capacitance, $C \equiv$ reciprocal of spring stiffness, $1/k$

8.2 First-order Systems

A first-order system is one which contains only *one* of the two possible types of energy storage element, in combination with the associated energy dissipative element. For the elements so far considered, the following combinations form first-order systems:

Mass-damper; spring-damper; inductor-resistor; and capacitor-resistor. The last two combinations we have already studied in detail when considering d.c. transients in Chapter 4. In this situation the systems (circuits) were subjected to a step input, by connecting to or disconnecting from a d.c. supply. In each case it was found that the system response followed an exponential law. Let us review the salient points regarding the C-R and L-R series circuits, and see how these results can be translated in terms of the response of analogous mechanical systems.

■ C-R series circuit and spring-damper system ■

Figure 8.1 shows a C-R circuit that is subjected to a sudden step input disturbance when the switch is closed. In this case the capacitor will charge up until the voltage between its plates is equal to the applied voltage. The relevant equations for this system are as follows.

$$E = v_R + v_C \text{ volt}$$

$$\text{since } v_R = Ri = \frac{Rdq}{dt}; \text{ and } v_C = \frac{q}{C}$$

$$\text{then, } E = \frac{Rdq}{dt} + \frac{q}{C} \text{ volt} \dots\dots\dots [1]$$

The above equation is known as the differential or system equation for the C-R circuit. We have seen

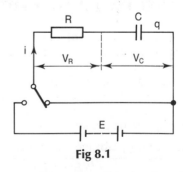

Fig 8.1

that its solution shows that the circuit p.d.s, charge, and current vary exponentially with time until the capacitor reaches its fully-charged state, such that:

$$v_C = E(1 - e^{-t/\tau}) \text{ volt; } i = I_o e^{-t/\tau} \text{ amp}$$

$$q = Q(1 - e^{-t/\tau}) \text{ coulomb; and } \tau = CR \text{ seconds}$$

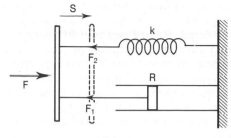

Fig 8.2

The spring-damper arrangement shown in Fig. 8.2 is subjected to a step input by suddenly applying a force F to it. The spring has a stiffness of k newton/metre, and the system will displace until the reaction force of the spring (F_2) is equal to the applied force. The damper has a damping coefficient of R newton per metre/second, and will exert a reaction force (F_1) only whilst the system is in motion. This may be compared to a resistor producing a p.d. only whilst charge is moving through it, i.e. whilst carrying current. Combining the forces in this mechanical system results in the following differential equation:

$$F = F_1 + F_2 \text{ newton}$$

where $F_1 = Rv = \dfrac{Rds}{dt}$; and $F_2 = ks$

so, $F = \dfrac{Rds}{dt} + ks$ newton .... [2]

Comparing equ [1] and [2] we can conclude that the displacement, s, of the system will increase, and the velocity, ds/dt, will decrease exponentially with time. Thus:

$$s = \frac{F}{k}(1 - e^{-t/\tau}) \text{ metre,}$$

where $\dfrac{F}{k}$ = final displacement

$$v = \frac{F}{R}e^{-t/\tau} \text{ metre/second,}$$

where $\dfrac{F}{R}$ = initial velocity

and system time constant,

$$\tau = \frac{R}{k} \text{ seconds}$$

Note: $C \equiv \dfrac{1}{k}$ or $k \equiv \dfrac{1}{C}$ as stated in Section 8.1

If the switch in the C-R circuit is now returned to its original position then the capacitor will discharge exponentially. Similarly, if the applied force is now removed from the spring, it will return to its normal length. This displacement will also be exponential in form.

■ *L-R* series circuit and mass-damper system ■

From Chapter 4 we know that when an L-R circuit is connected/disconnected to/from a d.c. supply, that the current will increase/decrease exponentially. The analogous mechanical system is the mass-damper subjected to a suddenly applied force, as shown in Fig. 8.3.

Since we have an analogous system we can conclude that the mass will accelerate until the reaction force of the damper causes the speed to stabilise at some constant value. The reason for this particular conclusion is that electrical current

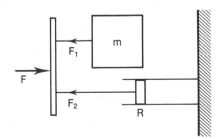

Fig 8.3

is analogous to velocity, and in the *L-R* circuit the current builds up to some steady value limited by the circuit resistance. The system differential equation for the mass-damper will be as follows:

$$F = F_1 + F_2 \text{ newton}$$

where F_1 = mass × acceleration

$$= ma \text{ newton} = m\frac{dv}{dt} \text{ newton}$$

and $F_2 = Rv$ newton

therefore, $F = m\dfrac{dv}{dt} + Rv$ newton

Hence, $v = \dfrac{F}{R}(1 - e^{-t/(\tau)}$ metre/second;

and $\tau = \dfrac{m}{R}$ seconds

where F/R is the final velocity.

The system equation for the *L-R* circuit is:

$$V = L\frac{di}{dt} + Ri \text{ volt}$$

The concept of first-order systems is not confined to the electrical and linear mechanical examples so far considered. Rotary mechanical systems, involving angular acceleration (α), angular velocity (ω), angular displacement (θ) and applied torque (T), will react in a similar manner. So too will the equivalent hydraulic and thermal systems. Thus the behaviour of *any* first-order system in response to a step input disturbance may be summarised as follows.

System equation of the form

$$Y = A\frac{dx}{dt} + Bx$$

the system variable, x, will respond exponentially such that

$$x = \frac{Y}{B}(1 - e^{-t/\tau}) \text{ for exponential growth}$$

$$x = \frac{Y}{B}e^{-t/\tau} \text{ for exponential decay;}$$

and $\tau = \dfrac{A}{B}$ seconds

■ Worked Example 8.1 ■

Q A spring-damper system as shown in Fig. 8.4 is subjected to a force of 25 N. The spring stiffness is 200 N/m and the damper coefficient is 600 Ns/m. If this force is suddenly removed, determine (a) the system time constant, (b) the time taken for the system to move 50 mm from its initial starting point, and (c) the time taken for the spring to return to its 'normal' uncompressed length.

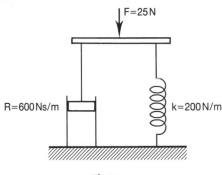

Fig 8.4

A

$F = 25$ N; $k = 200$ N/m; $R = 600$ Ns/m

(a)
$$\tau = \frac{R}{k} \text{ seconds} = \frac{600}{200}$$

therefore $\tau = 3$ s **Ans**

(b) In order to determine details of how the spring will return to its unstressed length, we firstly have to determine the distance the applied force will initially displace it.
Initial displacement,

$$s_i = \frac{F}{k} \text{ metre} = \frac{25}{200} = 125 \text{ mm}$$

Comparing this system with its electrical analogue, we can say that this displacement corresponds to the initial charge placed on a capacitor. When the force is released it will be equivalent to

starting the discharge process for the capacitor. Thus the displacement for which the time is required to be calculated involves a decaying exponential function.

$$s = s_i e^{-t/\tau} \text{ metre}$$

$$\frac{s}{s_i} = e^{-t/\tau}$$

$$-\frac{t}{\tau} = \ln \frac{s}{s_i}$$

$$-\frac{t}{3} = \ln \frac{50}{125} = -0.9163$$

hence $t = 2.75$ s **Ans**

(c) The system will take approximately five time constants to reach its new steady state, so time taken for spring to unstretch = 15 s **Ans**

8.3 Ideal Second-order Systems

A second-order system is one which contains *both* forms of energy storage element (one for 'KE' and one for 'PE'). Let us firstly consider an ideal second-order system. This is one in which there is no loss or dissipation of energy, hence the energy dissipative element is absent. Such an ideal is

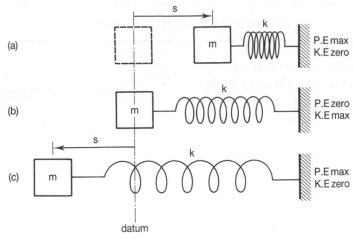

Fig 8.5

impossible to achieve in practice since it would result in 'perpetual motion'. However, we need to study the behaviour of such a system in order to appreciate the behaviour of a 'real' second-order system. Consider a mass-spring system as in Fig. 8.5, which is free from all damping effects, such as friction etc. Figure 8.5(a) illustrates the setting of the initial conditions, whereby the system is given an initial displacement by the application of a force.

Under these initial conditions the system displacement will be at its maximum value, and the energy input to the system will be stored as PE in the spring ($0.5\ ks^2$ joule). No energy will be stored in the mass at this stage since it will be stationary. At some time $t = 0$, the system is released by removing the applied force. The spring will assert itself and drive the mass back towards its normal datum position. In so doing the spring gives up energy to the system. Since the mass is being accelerated, then it is gaining KE ($0.5\ mv^2$ joule). This energy can only have been obtained from the spring. Thus, the PE in the spring is being transferred into KE in the mass.

Figure 8.5(b) shows the instant in time where the system has reached its normal datum point, (i.e. zero displacement). At this time the spring will have transferred all of its energy to the mass, so the mass will have reached its maximum velocity. The inertia of the mass will now cause it to continue beyond the datum point. As it does so the spring will begin to stretch. Thus the mass will now be transferring energy back to the spring.

Figure 8.5(c) shows the instant that the mass comes to rest, at the system maximum displacement on the other side of the datum. At this point all of the energy will once more be stored in the spring. The whole sequence will then be repeated, and the system will oscillate back and forth with a constant amplitude of s metre. Since no energy is lost from the system, then these oscillations would continue indefinitely—perpetual motion! In exactly the same way as was used in the first-order systems, we can deduce the system differential equation for this system by equating the forces involved, as follows.

$$\text{Inertia force due to the mass} = ma$$

$$= m\frac{dv}{dt}\text{ newton}$$

and stiffness force due to spring $= ks$ newton

$$\text{therefore, } F = m\frac{dv}{dt} + ks\text{ newton}$$

and since $v = ds/dt$; then $dv/dt = d^2s/dt^2$

$$\text{so, } F = m\frac{d^2s}{dt^2} + ks\text{ newton}$$

and at time $t = 0$, $F = 0$, so equation becomes:

$$0 = m\frac{d^2s}{dt^2} + ks$$

The above equation describes simple harmonic motion, and the solution to such an equation is a cosine waveform. This means that a graph of displacement will be a cosine waveform, as shown in Fig. 8.6. If the mass is large and the spring

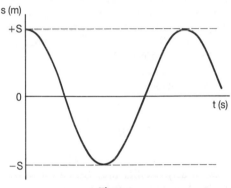

Fig 8.6

'soft', then the frequency of oscillation will be relatively low. On the other hand, a 'stiff' spring and small mass would result in a relatively high frequency of oscillation. In general we can conclude that the frequency of oscillation is in some way directly proportional to the spring stiffness k, and inversely proportional to the mass m. This frequency of oscillation is known as the natural frequency, f_n, for the system, and it may be calculated from:

$$\omega_n = \sqrt{\frac{k}{m}} \text{ rad/s} \qquad (8.1)$$

$$\text{so,} f_n = \frac{1}{2\pi}\sqrt{\frac{k}{m}} \text{ hertz} \qquad (8.2)$$

It may occur to you that these equations seem vaguely familiar. That is as it should be, since the equivalent electrical system is the *L-C* circuit shown in Fig. 8.7. In this circuit, the capacitor

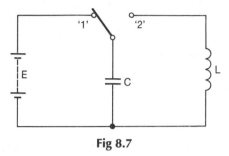

Fig 8.7

would initially be charged with the switch in position '1'. When the switch is moved to position '2' the energy stored in C will drive a current through L, thus transferring energy to the magnetic field of the inductor. The result will be a continuous oscillation of current and energy between the capacitor and inductor (assuming zero resistance). The frequency of oscillation would be:

$$\omega_n = \frac{1}{\sqrt{LC}} \text{ rad/s;} \quad \text{or} f_n = \frac{1}{2\pi\sqrt{LC}} \text{ hertz}$$

You should recognise the above equations as those for the resonant frequency for a series *L-C* circuit, or, for a *parallel* resonant circuit where the *resistance is zero*. Bearing in mind that according to our analogies $C \equiv 1/k$, then it may be seen that the resonant or natural frequency equation for the electrical circuit corresponds directly with that for the mechanical system shown.

8.4 Practical Second-Order Systems

Even if a damper element is not included in a mechanical system, there will always be some unavoidable damping effect due to friction. Similarly, in an electrical circuit, the connecting wires and inductors will have some resistance. Since any form of damping dissipates energy, then the oscillations induced into a practical second-order system must die away with time. Consider a practical mechanical second-order system that is given an initial displacement, and is then released.

The system is represented in Fig. 8.8. The reaction forces of the three elements are shown as F_1 to F_3, and the system differential equation will be:

$$F = F_1 + F_2 + F_3 \text{ newton}$$

$$\underbrace{F}_{\substack{\text{step input} \\ \text{disturbance}}} = \underbrace{m\frac{d^2s}{dt^2}}_{\substack{\text{'inertia'} \\ \text{term}}} + \underbrace{\frac{Rds}{dt}}_{\substack{\text{'damping'} \\ \text{term}}} + \underbrace{ks}_{\substack{\text{'stiffness'} \\ \text{term}}}$$

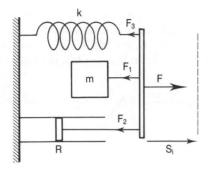

Fig 8.8

The solution of second-order differential equations is beyond the scope of the mathematics required for the course of study being undertaken, so the general solution is shown below.

$$s = Ae^{-\alpha t} \sin \omega_d t \text{ metre}$$

Analysing this equation shows that there is an oscillatory component ($\sin \omega_d t$), the amplitude of which ($Ae^{-\alpha t}$) is a decaying exponential. Thus the system will oscillate at a frequency of ω_d rad/s ($f_d = \omega_d/2\pi$ hertz), and these oscillations will decay exponentially with time. The response of the system, when the force F is removed, is shown in Fig. 8.9.

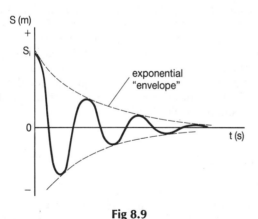

Fig 8.9

The term α in the decay component of the above equation is a direct function of the system damping. Thus the greater the damping, the more rapidly will the oscillations die away. Indeed, if the damping is increased sufficiently, a point will be

reached where the oscillations are just prevented from occurring. This degree of damping is known as critical damping R_{crit}. Under this condition the response of the system will be approximately exponential.

Although the system will have a specific value of damping coefficient R, the overall damping is normally referred to in terms of the system damping ratio ζ (Greek letter zeta). Damping ratio is defined as the ratio of the damping present to that required to just prevent oscillations. In the form of an equation this is:

$$\zeta = \frac{R}{R_{crit}} \qquad (8.3)$$

The degree of damping present also affects the frequency of any oscillations. Since the system contains both mass and spring stiffness, then the system will have a natural frequency of oscillation, f_n. As the damping is increased so the frequency of oscillation is decreased, so the damped frequency f_d is always less than the natural frequency. The relationship is given by:

$$f_d = f_n\sqrt{1 - \zeta^2} \text{ hertz} \qquad (8.4)$$

The equivalent electrical circuit is shown in Fig.

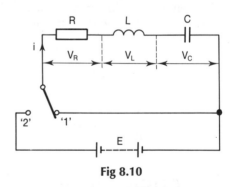

Fig 8.10

8.10, and the relevant system equations are listed below.

$$E = v_L + v_R + v_C \text{ volt}$$

$$\text{so, } E = L\frac{d^2q}{dt^2} + R\frac{dq}{dt} + \frac{q}{C} \text{ volt}$$

$$\omega_n = \frac{1}{\sqrt{LC}} \text{ rad/s}; f_n = \frac{1}{2\pi\sqrt{LC}} \text{ hertz}$$

$$\omega_d = \omega_n \sqrt{1 - \zeta^2} \text{ rad/s;}$$
$$f_d = f_n \sqrt{1 - \zeta^2} \text{ hertz}$$

In this circuit, when the switch is moved from position '1' to '2' the charge in the system will increase from zero to its final steady value, whilst the circuit current will decrease from its initial value, to zero. The graphs of these variations are illustrated in Fig. 8.11. Returning the switch to

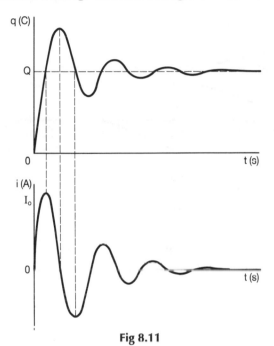

Fig 8.11

position '1' would be equivalent to releasing the mass-spring-damper as previously described.

The effect of damping ratio ζ on the response of *any* second-order system can be determined from a set of generalised response curves, an example of which is shown in Fig. 8.12. An alternative approach is to use a generalised form of the system differential equation. Consider the above electrical system:

$$E = L\ddot{q} + R\dot{q} + q/C$$

where $\ddot{q} = d^2q/dt^2$ and $\dot{q} = dq/dt$

and dividing through by the coefficient L:

$$\frac{E}{L} = \ddot{q} + \frac{R\dot{q}}{L} + \frac{q}{LC}$$

The coefficient of the last term is $1/LC = \omega_n^2$. It is found that the coefficient of the second term, $R/L = 2\zeta\omega_n$. Thus, if any second-order system has an equation of the form

$$Y = a\ddot{x} + b\dot{x} + cx$$

$$\text{then } \frac{Y}{a} = \ddot{x} + \frac{b\dot{x}}{a} + \frac{cx}{a}$$

$$\text{and } \frac{Y}{a} = \ddot{x} + 2\zeta\omega_n\dot{x} + \omega_n^2 x \qquad (8.5)$$

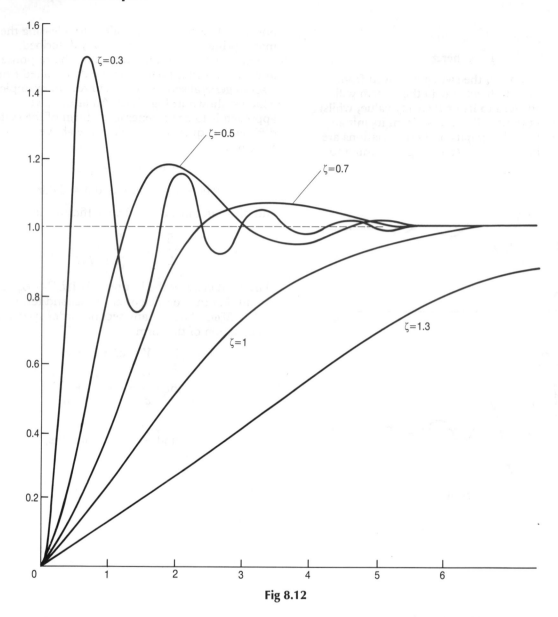

Fig 8.12

■ **Worked Example 8.2** ■

$\boxed{Q}$ A linear mechanical system has a total mass of 15 kg, an effective spring stiffness of 10 N/m, and a damping coefficient of 20 Ns/m. If this system is subjected to a sudden applied force of 25 N, write down the system differential equation and hence determine (a) the natural frequency, and (b) the system damping ratio.

$\boxed{A}$ ——————————————————————————————

$m = 15$ kg; $k = 10$ N/m; $R = 20$ Ns/m; $F = 25$ N

$\qquad F = m\ddot{s} + R\dot{s} + ks$ newton

$\qquad$ so $25 = 15\ddot{s} + 20\dot{s} + 10s$ **Ans**

(a)
$\qquad$ hence, $\dfrac{25}{15} = \ddot{s} + \dfrac{20\dot{s}}{15} + \dfrac{10s}{15}$

$\qquad$ and comparing coefficients with the general equation (8.5):

$\omega_n = \sqrt{\dfrac{10}{15}} = 0.816$ rads/s

so $f_n = \dfrac{0.816}{2\pi} = 0.13$ Hz **Ans**

(b)
$\qquad 2\zeta\omega_n = \dfrac{20}{15} = 1.3333$

therefore $\zeta_n = \dfrac{1.333}{2 \times 0.816} = 0.817$ **Ans**

8.6 Driven Damped Systems ——————————————————

A driving or forcing function input may have a waveshape that is sinusoidal, rectangular, triangular, or indeed any other waveshape. For mechanical systems, such a forcing function may be the result of vibration. In this case the waveshape will be approximately sinusoidal. Figure 8.13 represents a mechanical system subjected to a sinusoidal forcing function.

This system will have a natural frequency of oscillation f_n hertz, or ω_n rad/s. Provided that the forcing input has a frequency ω that is very much less than the system ω_n, then the mass in the system will closely follow the forcing input motion. This may easily be confirmed by the following simple test. Hold one end of a reasonably 'soft' spring which has a mass suspended from the other end. Now move your hand up and down very slowly, and you should find that the mass and

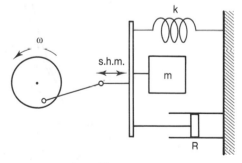

Fig 8.13

spring move in sympathy with the input displacement. As the input forcing frequency is increased (gradually increase the speed of the up-and-down movement of your hand) it will be found that the displacement of the mass starts to become out-of-step with the input. In addition it will be found that the amplitude of mass displacement also increases. At a particular input frequency it will be found that the displacement of the mass is 90° out-of-step with the input, and its amplitude will be considerably greater than the input displacement. Take care with your test rig, since under this condition the mass may well detach itself from the spring and do some damage! This effect occurs at the resonant frequency f_o for

the system. If the input frequency is increased further, then the amplitude of the oscillations of the mass will begin to decrease, and its angle of lag will continue to increase. There will come a point when the inertia of the mass is such that the mass cannot keep up with the driving function, and its displacement becomes virtually zero. The relationship between the system displacement to the frequency of the forcing function is shown in Fig. 8.14, and the angular relationship between

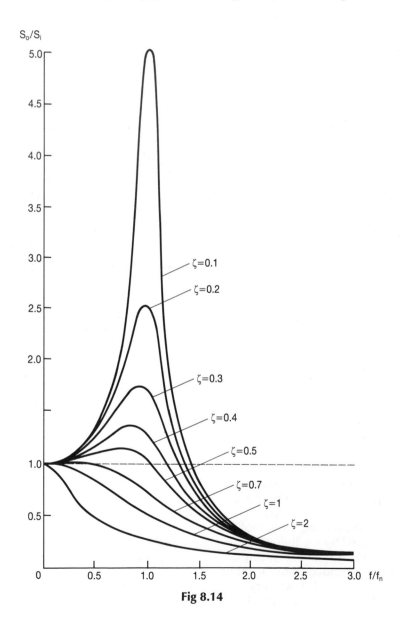

Fig 8.14

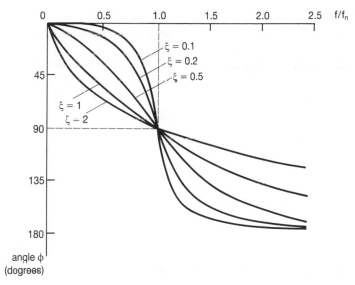

Fig 8.15

input and mass displacement is shown in Fig. 8.15. Note that both of these diagrams are 'generalised' (apply to any system) by making the frequency axes the ratio f/f_n in each case, and the vertical axis of Fig. 8.14 the amplitude ratio s_o/s_i. In the latter case, s_i represents the input displacement, and s_o represents the system displacement. Thus at very low input frequencies, $s_o = s_i$ (mass in step with input), so $s_o/s_i = 1$.

From Fig. 8.14 it is apparent that if the system is only lightly damped then the displacement and velocity of the mass can be very large at the resonant frequency. This could well lead to structural damage, and for this reason the resonant condition is normally avoided in mechanical systems. If inputs at or near the resonant frequency are unavoidable, then the system damping must be increased. This will have the effect of reducing the amplitude of oscillations. From the amplitude/frequency response curve it may be seen that if $\zeta \geqslant 0.7$, then the displacement of the mass will not exceed that of the input. Also, at frequencies in excess of $1.5f_o$ the displacement of the mass becomes progressively less and less. A classic example of these effects may be experienced when driving a car in which the roadwheels are out of balance. In this case it will be found that at one particular speed the vibrations due to the imbalance become very noticeable. If the speed is either increased or decreased from this value it will be found that the vibrations become much weaker, or even disappear altogether.

8.7 Control System Strategies

In Volume 1 we briefly considered the problem of controlling the position of a large radio telescope antenna. The initial solution to this problem involved the use of a very simple closed-loop control system, the diagram for which is shown in Fig. 8.16. It will be helpful at this stage to review the manner in which this system operates. This is summarised by describing the function of each of the system elements, as follows.

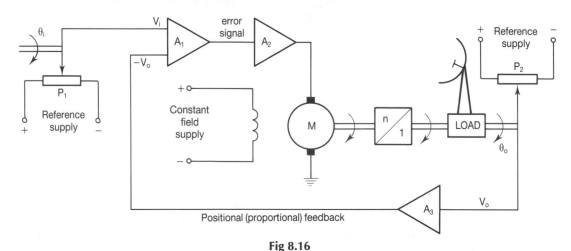

Fig 8.16

■ Proportional control ■

This control strategy is called proportional control because the feedback signal is directly proportional to the angular position of the output.

■ P_1 and P_2 ■

These are the input and output potentiometers (transducers) which provide the electrical signals (V_i and V_o) that are proportional to the input or demanded displacement and the actual output displacement, θ_i and θ_o respectively.

■ A_1 ■

This is the error detector, in the form of a summing operational amplifier. The output of this device will be the error signal, equal to $V_i - V_o$.

■ A_2 ■

This is a power or servo amplifier which provides a much amplified error signal to the drive motor.

■ M ■

This is a separately excited d.c. driving motor.

■ n ■

This is a step-down gearbox to provide the large torque required by the load.

■ A_3 ■

This is an inverting operational amplifier to ensure that the feedback signal V_o is in opposition to the input V_i. This ensures that negative feedback is applied.

In addition to the mass of the load are the masses of motor armature, drive shafts, gearbox etc. Therefore the total mass, and hence inertia, of the system will be large. The 'stiffness' of the system is contributed by the drive shafts and gearbox, and the amplification in the system will have a similar effect to mechanical stiffness. In order to minimise any steady state error the system gain (amplification) needs to be high. On the other hand, the friction in the system will be minimal due to the use of high quality bearings, lubrication etc. The whole system will therefore be a severely underdamped second-order system which, due to the high gain, will be very responive to demanded changes. Thus, when a step demand input is applied, the system will respond rapidly with consequent violent oscillations of the antenna about its new position before finally settling. This oscillatory response is obviously undesirable, especially since large masses are involved. The solution is to slow down the system by increasing the damping ratio.

This effect could be achieved by increasing the friction in the system, perhaps by the use of some form of braking. However, increasing viscous friction is to be avoided because it wastes energy, it increases steady state error, and it was initially designed to be minimal anyway! Thus the slowing down affect of viscous friction needs to be simulated, without the other adverse affects. This may be achieved by means of derivative feedback as follows.

■ Derivative control ■

The original positional (proportional) feedback must be retained since this forms the basic closed-loop control which ensures that the system is error-actuated. The derivative feedback may be achieved by several means, one of which is by the use of a tachogenerator mounted on the motor driveshaft as shown in Fig. 8.17.

The output of the tachogenerator is directly proportional to the speed at which it rotates, so it is proportional to the rate of change of output position, i.e. $V_d \propto d\theta_o/dt$, so the tacho voltage is the derivative of the output position. This feedback signal is also of the opposite polarity to the demand signal, so the total error signal will now be given by $V_i - (V_o + V_d)$ volt. The major advantage of tachogenerator feedback is that it will be at its greatest when the output shaft is travelling at its fastest (through the oscillatory 'cross-over' points on the response graph), and will be zero when the shaft is stationary. This means that the damping effect of this feedback is present only when it is required and will, unlike friction, not affect the

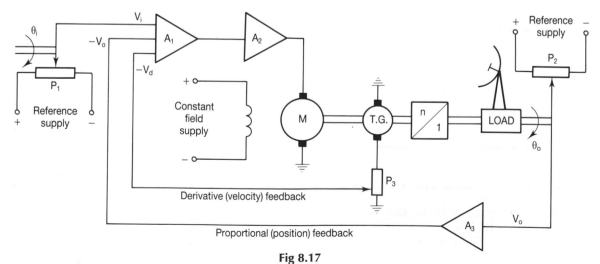

Fig 8.17

steady state error in the shaft position. In addition, the degree of damping produced can very simply be adjusted by means of the associated potentiometer, P_3.

Although the use of a tachogenerator would seem to be the ideal solution to the problem of oscillatory instability in a positional control system, it can have an adverse affect in a velocity control system, or a positional control system subjected to a ramp input demand.

■ Velocity lag ■

A ramp input is simply a step *velocity* input, as illustrated in Figs. 8.18(a) and (b). This type of

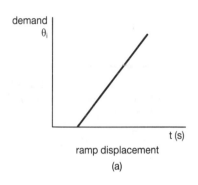

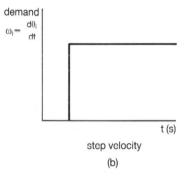

ramp displacement

(a)

step velocity

(b)

Fig 8.18

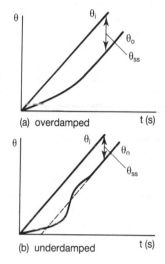

(a) overdamped

(b) underdamped

Fig 8.19

the error signal to maintain this velocity there will be an angular misalignment between the actual instantaneous position of the antenna and the desired position. This means that the antenna angular position will be 'lagging behind' the target. This angular difference is called the velocity lag, and from Figs. 8.19(a) and (b) it may be seen that velocity lag (steady state error) increases with damping. Hence if a tachogenerator is used, the damping signal that it produces whilst rotating will increase the velocity lag. This problem may be avoided if the velocity feedback is produced by passing the error signal through either a passive C-R differentiating network, or a differentiating operational amplifier. The differentiator and the error detector usually form parts of an element called a controller. Thus a system that employs both proportional and derivative feedback is said to have a two-term controller.

input would apply to a positional control system when it is required to rotate at some constant velocity. For example, this would apply to a radar antenna when it is required for the atenna to lock on to and follow a 'target'. Since the system has to be error-actuated, then to keep the motor driving there has to be *some* difference between θ_o and θ_i, to provide the necessary error signal. Consider the antenna rotating at the appropriate speed to match that of the 'target' ($\omega_o = \omega_i$). In order to provide

▪ Integral control ▪

The differentiating network mentioned above is satisfactory for low-power systems in which the load on the motor is relatively small and constant. However, where there is significant 'stiction' (static or coulomb friction, independent of velocity) in the system, then steady state errors can still be significant, since stiction increases the deadband. As the load approaches the desired alignment position ($\theta_0 = \theta_i$), the error voltage becomes progressively smaller. Figure 8.20 illustrates this point and shows the deadband. In this context, the dotted lines represent the minimum error voltage required to overcome the stiction. Thus, due to its inertia, the load could come to rest at any position between points (a) and (b). The system deadband could be reduced by increasing the system gain, but too much gain tends to produce instability, in the form of oscillations. The solution to the problem is the use of integral feedback, either in the form of a passive *C-R* integrating network or an integrating operational amplifier. In either case, a p.d. is built

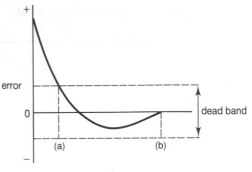

Fig 8.20

up across the capacitor whilst the load is approaching alignment. This p.d. is sufficient to supplement the error signal to the extent that sufficient error voltage will now be present to overcome the deadband. With careful adjustment it is possible to eliminate the steady state error. The arrangement for a three-term or PID controller is shown in Fig. 8.21.

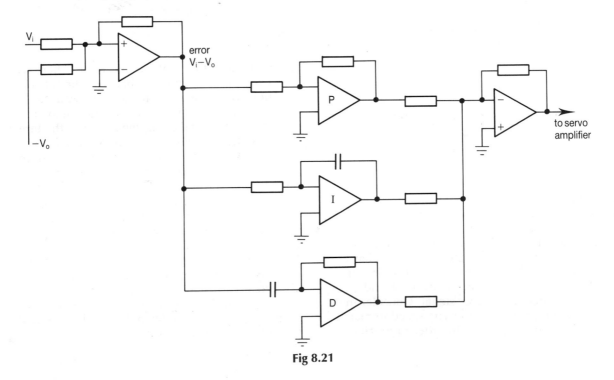

Fig 8.21

8.8 Computer Control of Systems

Increasingly the control of modern systems is achieved by microprocessors and microcomputers. One example is the microprocessor in a washing machine, which determines the sequence and timing of the various operations. In order to achieve this the processor acts on information provided by sensors (transducers) that monitor such variables as water level, temperature, drum speed, programme selection, etc. An example of the use of a computer would be the control of a manufacturing process, which could involve the control of, and response to, a large number of variables, including material position and availability, temperature, flowrate, pressure, tool position, velocity, safety features, etc. The information regarding all the variables will also be supplied by appropriate transducers. A block diagram for a typical system is shown in Fig. 8.22.

The variables to be sensed and controlled are analogue quantities, whereas the computer or microprocessor is a digital device. For this reason the information fed back from the transducers has to be converted into binary data which can be processed by the computer. This is achieved by a device known as an analogue to digital converter (ADC). Similarly, the control signals generated by the computer have to be converted from digital to analogue signals. This is achieved by a digital to analogue converter (DAC).

The control strategies described in the previous section may be achieved by a computer or microprocessor. The manner in which specific control actions are achieved will not be dealt with here, since such details form part of the modules (units) covering computer applications and microelectronic systems.

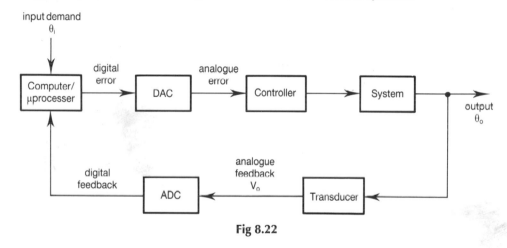

Fig 8.22

Assignment Questions

1 Deduce the electrical analogues for torque, inertia, angular velocity, angular displacement, and a torsion bar.

2 A spring of stiffness 1.5 kN/m is attached to a mass of 20 kg. What will be the natural frequency of oscillation of this system when disturbed?

3 A mass-spring system has a natural frequency of 1.5 Hz. If the mass is 4 kg determine the value of spring stiffness.

4 A mass-spring system has a natural frequency of 0.5 Hz. If it is then damped by a device having a damping ratio of 0.65, determine the damped frequency of oscillation.

5 Explain the difference between first-order and second-order systems, quoting examples of each.

6 A mass of 63 kg is attached to a damper of coefficient 30 Ns/m. This system is subjected to a suddenly applied force of 100 N. Write down the system differential equation and hence determine (a) the initial acceleration, (b) the time constant, and (c) the final velocity.

7 A system comprises a 2.4 kg mass, a spring of stiffness 9.6 N/m, a damper of coefficient 3.2 Ns/m, and is subjected to a step input force of 0.5 N. Determine the system differential equation and hence, or otherwise, calculate (a) the natural frequency, (b) the damping ratio, (c) the damped frequency, (d) the final displacement and (e) the value of damper coefficient that will just prevent oscillations.

8 Explain why resonance in mechanical systems is normally avoided, and if it cannot be avoided, how the adverse affects are minimised.

9 Explain what a tachogenerator is and why it may be employed in a positional control system.

10 (a) Explain what is meant by the term velocity lag, as applied to control systems, and how it may be reduced. (b) What is a three-term controller?

▪ 9 Information Transmission ▪

This chapter gives an overview of the principles of the transmission of digital information. The concepts of a d.c. and a.c. signals, and the terminology used is covered in Volume 1. On completion of this chapter you should be able to:

1] Compare analogue and digital transmission systems, outlining their relative advantages/disadvantages.

2] Describe the different forms of serial and parallel transmission of digital signals.

3] Describe the features of typical protocols.

4] Describe typical data network configurations.

5] Describe the features of local area and wide area networks (LANs and WANs)

9.1 Transmission Systems

Information is normally transmitted in the form of electrical signals, and the method of transmitting these electrical signals forms the basis for the different systems. The principal systems in use are television, radio, telephony, data communication and telegraphy.

Commercial television and radio broadcasting are 'one-way' communication systems, which rely on the propagation of electromagnetic radiations ('radio' waves) from one antenna to another. The methods by which this is achieved are dealt with in Chapter 10. Many long-distance telephone circuits also are routed via microwave radio-relay links or via communication satellites.

The majority of telephony communications are carried by the public switched telephone network (PSTN), the principal UK operator being British Telecom. Local circuits are normally routed over copper cables, though increasingly these are being converted to fibre optic cables. Eighty per cent of the long-distance circuits have already been converted to optic fibre cables. The latter have the following advantages over copper cables:

(i)   very wide bandwidth
(ii) low losses
(iii) high reliability and longevity
(iv) freedom from noise
(v)   lightweight and relatively inexpensive

The very wide bandwidth makes optical fibre cables particularly useful for digital signals, since these require much larger bandwidths than analogue signals. At present the majority of the UK telephone exchanges have been converted to use digital equipment for routing calls. It is anticipated that the conversion of the whole of the PSTN to digital operation will be completed in 1996. When this is accomplished there will be a completely digital transmission and switching network, known as the Integrated Service Digital Network (ISDN).

Telegraphy involves the transmission of information between teleprinters. This used to be via the Telex network, which operated in a similar manner to the PSTN. However the Telex service is now defunct, and communication between teleprinters is carried by the PSTN. To a very large extent the teleprinter service has been superseded by facsimile telegraphy or FAX. This system operates over the PSTN and can be used to transmit documents, photographs, diagrams etc. that have been converted into electrical signals by the FAX equipment.

Data communication refers to the transmission of information between computers or between computers and peripheral devices, such as VDUs, printers, disc drives, data loggers etc. Since computers are capable of handling only binary (two-state) signals, then the data to be transmitted is always in digital form.

9.2 Digital Signals

Whereas an analogue signal consists of a continuous waveform of varying amplitude, a digital signal consists of a train of voltage pulses. Thus digital signals are two-state or two-level signals; for example, ON–OFF, conducting–non conducting, 0 V–+5 V, etc. This type of signal is easily produced by transistors or diodes used as electronic switching devices. Figures 9.1 and 9.2

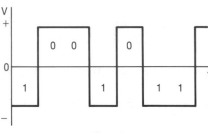

Fig 9.2

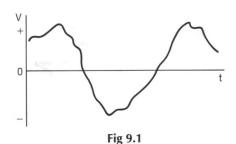

Fig 9.1

illustrate analogue and digital signals respectively.

The majority of the PSTN uses analogue transmission systems, but a digital signal cannot be directly transmitted via such a network. For this reason, a digital signal must be converted to a voice-frequency (vf) before transmission, and reconverted at the receiving end. The device that achieves this is called a modem (modulator/demodulator). Thus a digital signal may be transmitted in the same way as a speech signal, via an analogue system, by the use of a modem at each end of the system. When the ISDN is fully implemented, all information transmitted over the PSTN will be digital in form. The advantages of digital transmission compared to analogue techniques are:

(i)   Multiplexing is easier using digital time-division multiplex (tdm) rather than analogue frequency-division multiplex (fdm). The main reason for this is that the channel selection can be achieved by simple logic gates as opposed to expensive filter circuits.

(ii) The decrease of signal-to-noise ratio poses no problem since digital signals can be easily regenerated at regular intervals along the transmission line.

(iii) Data, FAX, speech etc. can be treated in exactly the same way for both transmission and switching.

(iv) Digital switching is simpler to achieve.

9.3 Serial and Parallel Transmission

For data communications, each character of the data to be transmitted has firstly to be converted into a digital (binary) signal, where a unique code is used to represent each character. The most common code is the ASCII code, in which eight bits (binary digits) are used—seven bits for the character, and the eighth for error-detection purposes (parity bit). Each group of eight bits is called a byte. When such data is to be transmitted, it may be either one bit at a time (serial transmission), or one byte at a time (parallel transmission).

> **Handshaking** is the procedure that allows two devices to communicate with each other in order to signal readiness to send/receive data, acknowledge receipt of data, synchronise clocks, etc.

■ Parallel transmission ■

In order to accomplish the transmission of a complete byte at a time there must be a line or conductor for each of the eight bits. There will also need to be provision for **handshaking** and other controlling signals. The *bus width* describes the number of parallel lines required in the connecting cable. Since this number will be at least 10 or 11, then parallel transmission is used only over very short distances, such as between the computer and a disc drive. However, parallel transmission enables very fast transfer of data. A parallel arrangement is illustrated in Fig. 9.3.

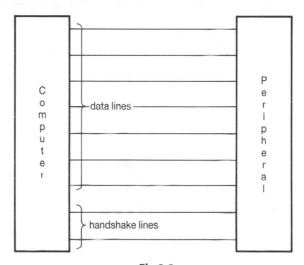

Fig 9.3

■ Serial transmission ■

When the data have to be transmitted over a distance in excess of about 1 metre then serial transmission is normally employed. This means that data are transferred one bit at a time, and the process therefore takes very much longer than parallel transmission. In its simplest form it requires very few lines, say one for sending and one for receiving data. However, the data within the computer are in parallel form (on the data bus), and this has to be converted into serial form before transmission. This process is accomplished by a device known as a UART (universal asynchronous receiver/transmitter) or an ACIA (asynchronous communications interface adaptor). The UART also converts incoming data from serial to parallel form. An example of a computer connected to a modem is illustrated in Fig. 9.4. Although not shown in this figure, some

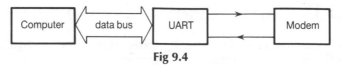

Fig 9.4

handshaking lines will also be required.

Within a given system a predetermined length of time must be allocated to one bit of data, and this time interval is defined and maintained by the computer *clock*. In order for the peripheral device to correctly interpret the serial data, it too needs to be controlled by an internal clock. These clocks, at either end of the transmission system, may either be synchronised or unsynchronised.

9.4 Asynchronous Data Transmission

Asynchronous (non synchronous) transmission is involved when in serial transmission, each character is sent as a complete but separate 'package'. The time interval between one character and the next is not fixed, yet the receiver must sample the incoming data bits at the correct instants and intervals of time. This element of the timing is controlled by the clock in the receiver, and the starting and stopping of this clock is controlled by *start* and *stop* bits transmitted with the character data. One method of achieving this is illustrated in Fig. 9.5. When the transmitter is not sending data the line voltage is maintained at the logic 1 (high) level. This is referred to as the 'line idle' state. When a character is about to be transmitted, the line is set to logic 0 (low) voltage level, for a time interval corresponding to one data bit. This is the start bit which enables (starts) the receiver clock. This is followed by the seven data bits plus the parity bit, and immediately followed

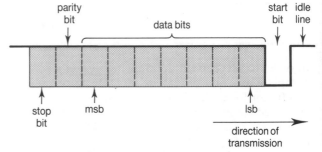

Fig 9.5

by the stop bit. The latter involves returning the line to the high state, which inhibits the receiver clock. This clock therefore needs to be synchronised with the computer clock only for the relatively short period required to transmit an individual character. For this reason the receiver clock need not be very stable.

9.5 Synchronous Data Transmission

With this system the use of start and stop bits is not required. The receiver clock runs continuously, and therefore needs to be both accurate and synchronised at all times with the computer clock. The other major difference is that bytes of data are transmitted continuously, without any breaks or gaps between them. Such a data stream is illustrated in Fig. 9.6. From this diagram it should be apparent that the receiver circuitry needs to be more sophisticated, so that it can distinguish one character byte from the next.

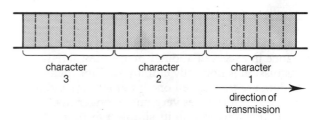

Fig 9.6

The receiver starts sampling the data as the first bit arrives. This is known as bit synchronisation.

The rate at which data is transmitted is given by the number of bits transmitted per second. The commonly adopted unit is the **baud**, where 1 baud is *roughly* equivalent to 1 bit/s. However, in a typical link a total of 11 bits is required to send only 7 bits of data. Thus for a line baud rate of 600, the usual data transfer rate will be only 382 bits/s.

Once this has been done the receiver identifies the group of bits that represent the character. This is called character synchronisation. Once these two synchronisations are established, the receiver can continue to correctly identify all the subsequent bytes transmitted to it.

9.6 Comparison of Asynchronous and Synchronous Transmission

An asynchronous system is much simpler and therefore less expensive one. The receiver clock may be relatively simple since it does not need to retain high accuracy over extended time intervals. However, such a free-running clock can give satisfactory performance only at relatively low bit rates, and the inclusion of start and stop bits also limits the maximum **baud** rate possible. The synchronised system can cope with much higher baud rates, but, because of its greater complexity, is more expensive. However, whichever of these transmission methods is used, a set of rules (the protocol) must exist so that the pin connections, logic conventions, error detection methods, etc. are known and adhered to regardless of the manufacturer of the devices used.

In this context, **duplex** refers to the ability to transmit and receive data simultaneously.

9.7 Protocols

There are a considerable number of protocols in existence, and we shall briefly consider two of the most widely used: the RS232C or CCITT V24, and the HDLC.

▪ RS232C/V24 ▪

This interface is the most widely used method of providing the serial transmission of data between computers and peripheral devices. The RS232C system may be used for the direct transmission of data up to 60 metre or so. For distances greater than this the use of a modem and telephone lines are more appropriate. The RS232C specification provides for the following signals.

(a) serial data comprising

(i) a primary channel providing full **duplex** data transfer.
(ii) a secondary channel also capable of full duplex operation.

(b) handshake signals.

(c) timing signals.

The RS232C may therefore be used in a variety of ways, including transmitting only (primary channel), receiving only (primary channel), half duplex, full duplex, and various combinations of these utilising both channels. This system is therefore very versatile and highly adaptable. The RS232C interface is usually achieved with a 25-way 'D' connector, but in practice, few systems involving computers make use of all of the available lines. More commonly the interface uses six signal lines and two ground (earth) connections, making eight lines in total. These use pins 1 to 7, and pin 20 on the 'D' connector, and their functions (at the computer end) are as follows:

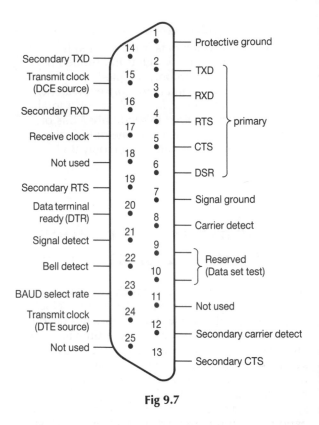

Fig 9.7

PROTECTIVE GROUND: Connected to the equipment frame or screening conductor.
TXD: Transmitted data output.
RXD: Received data input.
RTS: Request to send data (output). Peripheral to transmit when line is high.
CTS: Clear to send (input). Line high indicates peripheral ready.
DSR: Data set ready (input). High indicates handshaking completed.
SIGNAL GROUND: Common signal return path and reference level. Must NOT be connected to pin 1.
DTR: Data terminal ready (output). High indicates that peripheral should be connected to communication channel.

The complete pin-out diagram for the 'D' connector is shown in Fig. 9.7, which indicates all of the possible functions using this protocol.

9.8 High Level Data Link Control (HDLC)

This protocol was originally designed for two-way simultaneous operation between a 'primary' station and one or more 'secondary' stations, on a point-to-point or a multi-point link. The primary station controls the data flow in the link by authorising secondaries to transmit. There is no provision for secondaries to communicate with each other. The transmission of data is continuous, and is only interrupted when an error is detected. The acknowledgement of correct receipt of data is also transmitted on a continuous basis, in the opposite direction. The principle is illustrated in Fig. 9.8.

The three main features of HDLC are:

(a) The frame structure

(b) The elements of procedure

(c) The classes of procedure

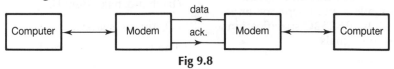

Fig 9.8

■ Frame structure ■

The frame structure is illustrated in Fig. 9.9. Flags are used to indicate the end of one frame and the start of the next, and the frames are transmitted in a continuous stream. If the transmitter has no information to send, it may transmit either continuous binary 'ones' or flags. Each frame consists of an 8-bit address field, an 8-bit control field, an information field of variable bit-length, and a 16-bit frame check sequence. The information field carries the message to be transmitted. The frame check sequence is the error detecting mechanism, and is applied to all fields of the frame except for the flags.

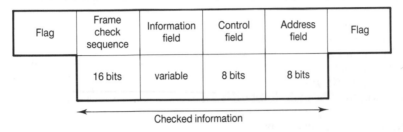

Flag	Frame check sequence	Information field	Control field	Address field	Flag
	16 bits	variable	8 bits	8 bits	

Checked information

Fig 9.9

■ Elements of procedure ■

These are concerned mainly with the use of the control field, which controls the operation of the link. This will involve the control of such things as which secondaries may transmit, and when; the acknowledgement (ACK) signals; and the number of frames that may be acknowledged by a single ACK signal, etc.

■ Class of procedure ■

There are two classes. In the unbalanced class, one station (the primary) has total responsibility for control of the link. Only the primary may issue commands and these always contain the address of the secondary for whom the command is destined. Responses by a secondary always contain the address of that secondary.

The characteristic of the balanced class is that stations at either end of a data link are *combined* stations, which can perform as both primary and secondary stations, as required.

9.10 Data Network Configurations (Topology)

A data network is a group of computers, terminals and associated peripheral equipment that are interconnected. The network may cover any area from a single room; a complete building; a site including a number of buildings; up to global communications. A given network may employ one or more types of network configuration (topology).

■ Bus network ■

This is a single communication circuit which is shared by all of the network devices; all of these being connected to this circuit. This topology is illustrated in Fig. 9.10, where the circular symbols represent the devices concerned. Each device would have a unique address. The main limitation to this topology is the maximum permitted length of the bus. For example, using a thick coaxial cable in the Ethernet system, the maximum length is about 500 m. The advantages of the bus network are the ease with which devices can be added to or removed from the network and the ease of locating cable faults.

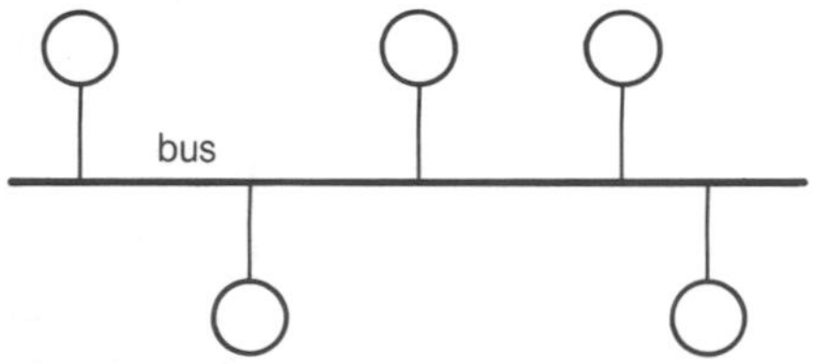

Fig 9.10

■ Ring network ■

The arrangement of a ring network is shown in Fig. 9.11, where each device (or node in the network) is connected to two others. Each node will normally have the facility to regenerate the signal, so the critical factor governing signal quality is the distance between a pair of nodes. However, this introduces some delay in transmitting a message around the ring. Even if the ring is broken at one point messages can still be propagated in the other direction around it. There are no routing problems since the data are available to all nodes, with the addressed node being able to copy the data.

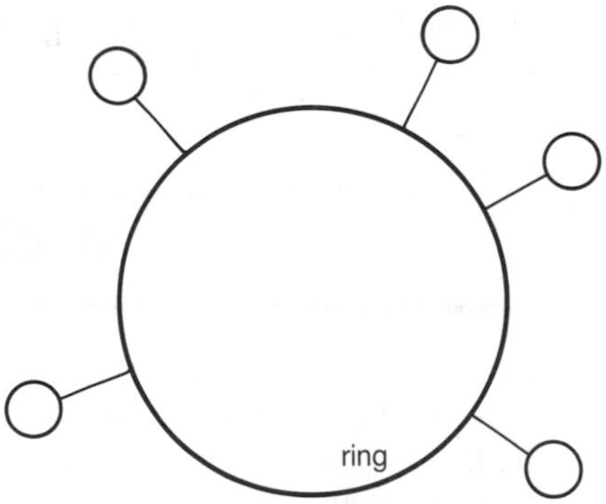

Fig 9.11

▪ Mesh network ▪

This is an extension of the ring topology, whereby some or all of the nodes are fully interconnected. The principle is illustrated in Figs. 9.12(a) and (b) This arrangement provides a large number of communication paths, with the advantage that any failure of a link or a node causes minimum problem. The disadvantage is the large amount of cabling required, and hence a larger cost. For this reason partially connected meshes are used in which only the most important nodes are fully interconnected.

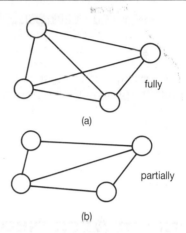

fully

(a)

partially

(b)

Fig 9.12

▪ Star network ▪

This topology is shown in Fig. 9.13. All messages have to be routed via the central switching device. This makes for a reliable system, provided that the switching device does not malfunction. In this event the whole network ceases to function. The other main disadvantage is that high data transfer rates cannot be achieved beause all messages are routed through the 'hub' of the system. This can be an economical installation where it is possible to utilise existing telephone wiring in a building.

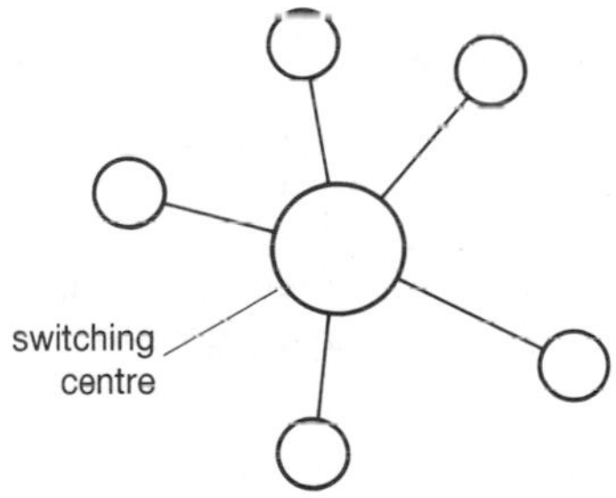

switching centre

Fig 9.13

■ Tree network ■

This is an extension of the bus topology, and is sometimes referred to as a hierarchical topology. As shown in Fig. 9.14 the main bus has spurs branching out from it. These spurs may also be split into further spurs. If this system is used to cover any significant area, then repeaters (regenerators) will be necessary at the various branching points.

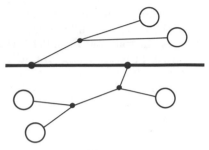

Fig 9.14

9.11 Local Area Network (LAN)

A local area network is simply a data network that provides the interconnection between computers, and between these computers and shared peripheral devices that are contained within a building or several buildings on a site. The limitation is that the distance between the extremities of the LAN is no more than about 2.5 km. The connecting cabling may be either coaxial or optic fibre. Since the bandwidth constraints of the PSTN do not apply, LANs may operate with very high bit rates; up to 100 Mbit/s where optic fibre is used. The network topology for a LAN may be any one (or combination of) the topologies already described. The physical layout of the building(s) will have a considerable affect in determining the actual combination of topologies chosen.

The advantages of a network are:

(a) A number of users can share common peripherals such as printers, fileservers, faxservers etc.

(b) A number of microcomputers can access one another and/or access a larger host computer.

(c) Programs and data may be stored anywhere in the network, but may be accessed by any user.

(d) The LAN may provide access to a wide area network via modems and other links.

9.12 Wide Area Network (WAN)

A wide area network is one which covers far greater distances than a LAN, up to and including global systems. For this reason the connections between stations will involve the use of analogue and digital telephone circuits, which in turn may involve the use of microwave links and satellite communications. Due to the cost of using telephone systems, multiplexing becomes an important feature in order to send as much data over the links, at as high a bit rate as possible. Two of the services offered by British Telecom are Kilostream and Megastream, and being digital systems the use of modems is not required. The Kilostream service offers bit rates of 2.4 kbit/s up to 64 kbit/s, whilst the Megastream service offers bit rates of 2.048 Mbit/s up to 140 Mbit/s.

Assignment Questions

1 Compare the relative advantages/ disadvantages of serial and parallel data transmission.

2 Compare the relative merits of synchronous and asynchronous data transmission.

3 Explain what is meant by the term *protocol*, and name two commonly used protocols.

4 Briefly explain the HDLC protocol. Include a description of the frame structure employed.

5 Explain the difference between a WAN and a LAN. Describe four topologies that may be used with LANs.

6 Compare the relative merits of Bus, Mesh, and Star networks.

7 Explain how digital data may be transmitted over an analogue system such as the PSTN.

8 Explain the advantages of digital transmission compared with analogue transmission methods.

9 Explain why 'raw' digital data cannot be transmitted directly over the PSTN.

■ 10 Modulation Techniques ■

The reasons for, and principles of amplitude and frequency modulation are covered in Volume 1. In this chapter we shall consider both procedures in more detail, and also consider the application of pulse modulation techniques. On completion you should be able to:

1] Explain the principles of various modulation techniques, and identify the type of modulation employed by the use of an oscilloscope.

2] Understand the terminology used and perform simple calculations for modulated waveforms.

3] Explain the operation of a simple demodulator.

4] Compare the characteristics of am and fm.

5] Explain how pulse modulation techniques are used to transmit analogue signals.

10.1 Amplitude Modulation (am)

In this form of modulation the instantaneous amplitude of a high-frequency sinusoidal *carrier* waveform is controlled by the instantaneous amplitude of a modulating *signal*. The resulting waveform is known as the *modulated carrier*. In practice, virtually all modulating signals will be complex waveforms, such as speech, music, pulse trains etc. The mathematical analysis of the resulting modulated carrier would be very complicated, so we shall consider the simplest case, which is a sinusoidal signal. Such a signal, the unmodulated carrier and the resulting modulated carrier are illustrated in Fig. 10.1. Since the modulated carrier to be transmitted is a complex wave, then it must consist of a number of sinewaves, of different frequencies and amplitudes,

that are effectively added together. In order to determine these so-called modulation products we need to analyse the waveforms, as follows:

Let modulating signal be
$$v_m = \hat{V}_m \sin \omega_m t \text{ volt}$$
and unmodulated carrier
$$v_c = \hat{V}_c \sin \omega_c t \text{ volt}$$

It is apparent from Fig. 10.1 that the amplitude of the modulated carrier varies in sympathy with the modulating signal, about its 'normal' amplitude of $\hat{V}_c$. Thus, the amplitude of the modulated carrier is $\hat{V}_c + \hat{V}_m \sin \omega_m t$, and its frequency remains unchanged. The expression for the modulated carrier is therefore:

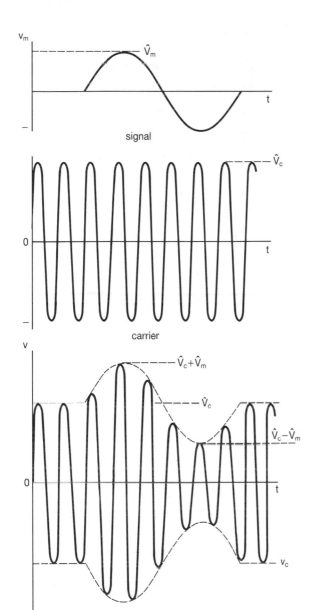

Fig 10.1

$$v = (\hat{V}_c + \hat{V}_m \sin \omega_m t) \sin \omega_c t \text{ volt} \qquad (10.1)$$

$$= \hat{V}_c \sin \omega_c t + \hat{V}_m \sin \omega_m t \sin \omega_c t$$

and expanding the second term in the above expression:

$$v = \hat{V}_c \sin \omega_c t + \frac{\hat{V}_m}{2} \cos(\omega_c - \omega_m)t$$

$$- \frac{\hat{V}_m}{2} \cos(\omega_c + \omega_m)t \text{ volt}$$

It may be seen that the above expression contains three voltage waveform components as follows.

(a) The original unmodulated carrier, of amplitude $\hat{V}_c$ volt, and frequency f_c hertz.

(b) A cosine wave of amplitude $\hat{V}_m/2$ volt, and frequency $(f_c - f_m)$ hertz.

(c) A negative cosine wave of amplitude $\hat{V}_m/2$ volt, and frequency $(f_c + f_m)$ hertz.

The frequencies $(f_c - f_m)$ and $(f_c + f_m)$ are known as the lower and upper side frequencies respectively. Figure 10.2 shows these three voltage waveforms and the resulting waveform when they are added together, i.e. the modulated carrier. A simpler way to illustrate the sidefrequencies etc. is shown in the frequency spectrum diagram of Fig. 10.3.

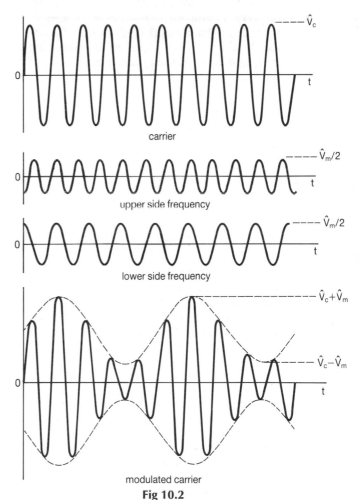

Fig 10.2

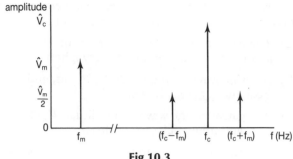

Fig 10.3

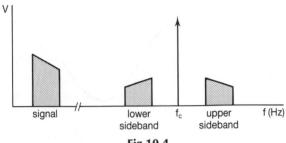

Fig 10.4

When the modulating signal is a complex waveform then each frequency component will produce a pair of side frequencies in the modulated carrier. Thus if there are N harmonics then there will be a total of $2N$ side frequencies, and the term *sidebands* is used to describe the lower and upper bands of frequencies produced by the modulation process. These are illustrated in Fig. 10.4, and from this diagram it can be appreciated that the total bandwidth required for the transmission system will be the difference between the highest upper sideband frequency and the lowest lower sideband frequency. Therefore the bandwidth can be calculated from:

$$\text{Bandwidth} = (f_c + f_m) - (f_c - f_m)$$

$$= 2f_m \text{ hertz} \qquad (10.2)$$

where f_m is the highest signal frequency component.

▪ **Worked Example 10.1** ▪

$\boxed{\text{Q}}$ **An rf carrier is amplitude modulated by a band of speech frequencies, 150 Hz to 4.5 kHz. Determine (a) the transmission bandwidth, and (b) the frequency components present in the transmitted wave if the carrier frequency is 506 kHz.**

$\boxed{\text{A}}$

(a) Bandwidth $= 2f_m$ hertz
           $= 2 \times 4500 = 9$ kHz **Ans**

(b)
      upper sideband $= (506000 + 150)$ to
                       $(506000 + 4500)$ Hz
                $= 506.15$ kHz to 510.5 kHz **Ans**

lower sideband $= (506000 - 4500)$ to
                 $(506000 - 150)$ Hz
          $= 501.5$ kHz to 505.85 kHz **Ans**

and the carrier $= 506$ kHz **Ans**

10.2 Modulation Index

The modulation index, m, is defined as the ratio $\hat{V}_m/\hat{V}_c$. From Fig. 10.3 it may be seen that:

maximum value of modulated carrier
$$= \hat{V}_c + \hat{V}_m \ldots \ldots [1]$$

minimum value of modulated carrier
$$= \hat{V}_c - \hat{V}_m \ldots \ldots [2]$$

Adding [1] and [2] yields
$$2\hat{V}_c = \text{max} + \text{min} \ldots \ldots [3]$$

and subtracting yields
$$2\hat{V}_m = \text{max} - \text{min} \ldots \ldots [4]$$

and dividing [4] by [3] we have

$$\frac{\hat{V}_m}{\hat{V}_c} = m = \frac{\text{max} - \text{min}}{\text{max} + \text{min}} \qquad (10.3)$$

The modulated carrier may now be expressed in terms of the modulation index by rewriting equation (10.1) as follows:

$$v = \hat{V}_c\left(1 + \frac{\hat{V}_m}{\hat{V}_c} \sin \omega_m t\right) \sin \omega_c t$$

hence, $v = \hat{V}_c(1 + m \sin \omega_m t) \sin \omega_c t$ volt $\qquad (10.4)$

■ Worked Example 10.2 ■

 The amplitude of an am waveform varies sinusoidally between 20 V and 2 V. Determine (a) the modulation index, and (b) the amplitudes of the modulating signal and the unmodulated carrier.

A

(a)

$$m = \frac{\text{max} - \text{min}}{\text{max} + \text{min}} = \frac{20 - 2}{20 + 2} = \frac{18}{22}$$

therefore, $m = \dfrac{9}{11} = 0.818$ or 81.8% **Ans**

(b) Since $m = \hat{V}_m/\hat{V}_c$; then $\hat{V}_m = 9$ V, and $\hat{V}_c = 11$ V **Ans**

If the amplitude of the signal equals that of the original carrier, $\hat{V}_m = \hat{V}_c$, then the modulation index will be 1, or 100%. In this case the amplitude of the modulated carrier will vary between $2\hat{V}_c$ and zero, as shown in Fig. 10.5(a). If, however, $\hat{V}_m > \hat{V}_c$, the result will be overmodulation ($m > 1$ or $m > 100\%$), as shown in Fig. 10.5(b). This diagram clearly shows that the envelope of the modulated carrier no longer reproduces the waveshape of the signal, so severe distortion has been introduced. For this reason the modulation index should not exceed 100%. For commercial broadcasting, the modulation index is normally limited to a maximum of 80% in order to ensure that overmodulation cannot occur.

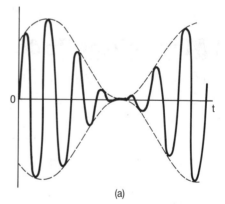

(a)

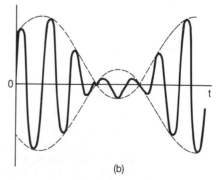

(b)

Fig 10.5

10.3 Power in the am Waveform

It has been shown that the modulated carrier contains three sinusoidal voltage components; the carrier of amplitude $\hat{V}_c$ volt and the two side frequencies, each of amplitude $\hat{V}_m/2$ volt. If the waveform is developed across a resistor of R ohm, then each component will contribute to the power dissipated. Converting the amplitudes to rms values, the power is given by:

$$P = \left(\frac{\hat{V}_c}{\sqrt{2}}\right)^2 \frac{1}{R} + \left(\frac{\hat{V}_m}{\sqrt{2}}\right)^2 \frac{1}{R} + \left(\frac{\hat{V}_m}{\sqrt{2}}\right)^2 \frac{1}{R} \text{ watt}$$

$$= \frac{\hat{V}_c^2}{2R} + \frac{\hat{V}_m^2}{4R}$$

$$= \frac{\hat{V}_c^2}{2R}\left(1 + \frac{\hat{V}_m^2}{2\hat{V}_c^2}\right)$$

and since $m = \hat{V}_m/\hat{V}_c$, then

$$P = \frac{\hat{V}_c^2}{2R}\left(1 + \frac{m^2}{2}\right) \text{ watt} \qquad (10.5)$$

Study of equation (10.5) will show that as the amplitude of the modulating signal is increased (m increased), so the power in the transmitted waveform increases. If there is no modulating signal ($m = 0$), then not surprisingly, the power contribution comes only from the carrier. For 100% modulation ($m = 1$), the power in the sidebands will be one third of the total power, i.e. one sixth of the total power in each sideband.

10.4 Demodulation

At the receiving end of the transmission system, the am waveform has to be demodulated ('decoded') in order to extract the signal information. This process is accomplished by a demodulator circuit, the basic principles of which are illustrated in Figs. 10.6 and 10.7.

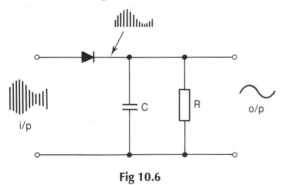

Fig 10.6

The envelope of the modulated waveform contains the signal information duplicated in both top and bottom halves. Since only one of these is required, the diode effectively 'passes' only the top half. This waveform is then applied to the *C-R*

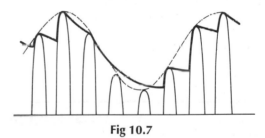

Fig 10.7

arrangement. The latter should bring to mind a smoothing capacitor arrangement used with a traditional rectifier circuit. Indeed, this is virtually what is needed, since the high-frequency carrier component is to be suppressed, whilst the relatively low-frequency signal component is required. In order to achieve this effect, the time constant of the circuit (CR seconds) needs to be long in comparison to the period of the carrier, but short compared to the period of the signal. The charge/discharge pattern of the *C-R* circuit is shown in more detail in Fig. 10.7, where it may be seen that a very close approximation to the original signal is produced.

10.5 Frequency Modulation (fm)

Frequency modulation involves the variation (deviation) of the instantaneous frequency of a sinusoidal carrier, in sympathy with the instantaneous amplitude of the modulating signal. The amplitude of the modulated carrier remains unchanged from its unmodulated form. As the modulating signal amplitude increases in its positive half-cycle, the frequency of the carrier is increased proportionately. In the other half-cycle of the signal, the carrier frequency is decreased. These effects are illustrated in Fig. 10.8 for a squarewave signal, and in Fig. 10.9 for a sinusoidal signal. The following terminology is used in fm systems.

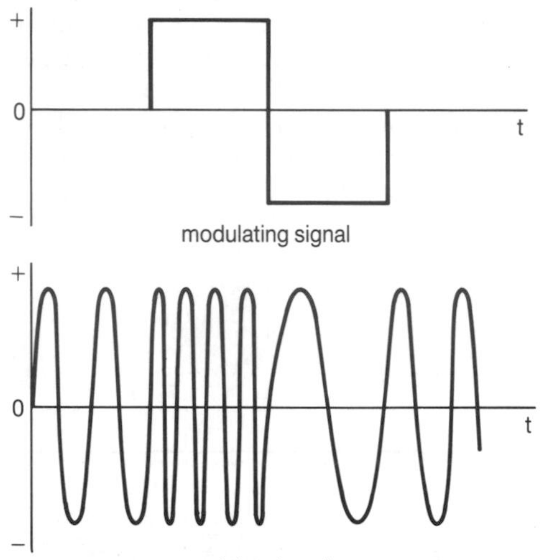

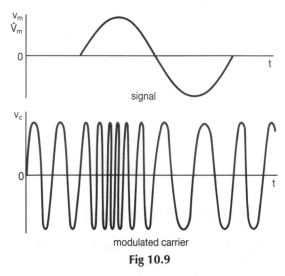

Fig 10.8

Fig 10.9

■ Frequency deviation, f_d ■

This is the difference between the highest frequency of the modulated carrier and its unmodulated frequency, in one cycle of modulation. The frequency deviation is directly proportional to the peak value of the signal.

■ Modulator sensitivity ■

This is the ratio of the frequency deviation to the amplitude of the modulating signal.

$$\text{i.e. sensitivity} = \frac{f_d}{\hat{V}_m} \text{ kilohertz/volt} \qquad (10.6)$$

■ Modulation index, m_f ■

This is the ratio of the frequency deviation to the frequency of the modulating signal.

$$m_f = \frac{f_d}{f_m} \qquad (10.7)$$

Also, m_f = the peak *phase* deviation of the carrier, measured in radian.

■ Frequency swing ■

This is the difference between maximum and minimum values of modulated carrier frequency. Generally this will be $2 \times f_d$ hertz.

■ Rated system deviation, f_{dr} ■

This is the maximum frequency deviation permitted in a given system. The bandwidth required for an fm wave is greater than the frequency swing, and it increases with the frequency deviation. For this reason some maximum value must be specified for the deviation allowed in a given system. For BBC fm broadcasts, the rated system deviation is 75 kHz.

■ Rated maximum modulating frequency, f_{mr} ■

This is the maximum modulating frequency that a given system can handle.

■ Deviation ratio, D ■

This is the ratio of the rated system deviation to the rated maximum modulating frequency.

$$D = \frac{f_{dr}}{f_{mr}} \qquad (10.8)$$

In order to put some of these terms and definitions into perspective, let us consider some numerical examples.

■ Worked Example 10.3 ■

 A 5 kHz sinewave of amplitude 1.5 V is used to frequency-modulate a 100 MHz carrier. If the modulator sensitivity is 20 kHz/V, determine (a) the frequency deviation, and (b) the modulation index.

$f_m = 5$ kHz; $\hat{V}_m = 1.5$ V; sens = 20 kHz/V;
$\qquad f_c = 100$ Mhz

(b)

$$m_f = \frac{f_d}{f_m} = \frac{30}{5} = 6 \textbf{ Ans}$$

(a)   sensitivity $= f_d/\hat{V}_m$, so $f_d = \hat{V}_m \times$ sens.

therefore $f_d = 1.5 \times 20 = 30$ kHz **Ans**

■ Worked Example 10.4 ■

 An fm system using a modulator of sensitivity 25 kHz/V has a rated deviation of 70 kHz. Calculate (a) the maximum permissible amplitude for the modulating signal, (b) the maximum value for the modulating signal frequency if the modulation index is to be restricted to 4, and (c) the resulting deviation ratio.

Sens. = 25 kHz/V; $f_{dr} = 70$ kHz; $m_f = 4$

(c)

$$D = \frac{f_{dr}}{f_{mr}} = \frac{70}{7.5} = 9.3 \textbf{ Ans}$$

(a)

$$\hat{V}_m = \frac{f_{dr}}{\text{sens.}} = \frac{70}{20} = 3.5 \text{ V } \textbf{Ans}$$

(b)

$$m_f = \frac{f_d}{f_m}; \text{ so } f_{mr} = \frac{f_d}{m_f} \text{ hertz}$$

therefore, $f_{mr} = \frac{30}{4} = 7.5$ kHz **Ans**

10.6 Frequencies Present in an fm Waveform

The level of mathematics required to carry out a rigorous analysis of a frequency-modulated wave is far beyond the scope of BTEC level 2/3 mathematics, so the following statements will need to be taken on trust.

The instantaneous value for an fm wave may be expressed as:

$$v = \hat{V}\{\sin \omega_c t \cos (m_f \cos \omega_m t)$$
$$- \cos \omega_c t \sin (m_f \cos \omega_m t)\}$$

and the terms $\cos(m_f \cos \omega_m t)$ and $\sin(m_f \cos \omega_m t)$ can be further expanded into the terms:

$\cos \omega_m t$, $\cos 2\omega_m t$, $\sin 2\omega_m t$, $\cos 3\omega_m t$, $\sin 3\omega_m t$, etc.

The fm wave therefore contains frequency components at the carrier frequency, f_c and an infinite number of side frequencies on either side of the carrier frequency. These side frequencies will be $(f_c \pm f_m)$, $(f_c \pm 2f_m)$, $(f_c \pm 3f_m)$, and so on. A plot of the amplitudes of the side frequency components, relative to that of the carrier, for side frequencies up to $(f_c \pm 9f_m)$ is shown in Fig. 10.10.

Considering Fig. 10.10, the following points should be noted.

1 The amplitude of the carrier and/or some of the side frequencies is zero at certain values of modulation index.

2 For small values of modulation index (say $m < 1$), the fm wave will occupy a bandwidth no greater than that required had the carrier been amplitude modulated. This is known as *narrow-band frequency modulation* (nbfm).

3 The larger the value of m_f the greater the bandwidth requirement for the transmission system

In general we can say that the bandwidth requirement for an fm transmission is given by the expression:

$$B = 2(f_d + f_m) \text{ hertz} \qquad (10.9)$$

where f_d is the maximum deviation and f_m is the maximum modulating frequency component.

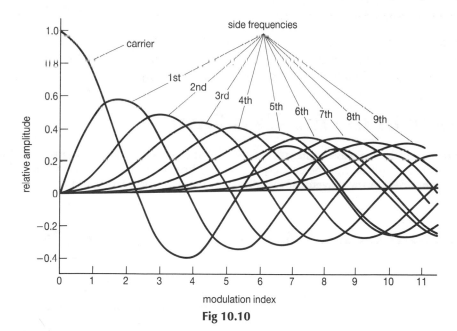

Fig 10.10

■ Worked Example 10.5 ■

A 15 kHz modulating signal is of such an amplitude that it produces a frequency deviation of 75 kHz. Determine the modulation index and the resulting bandwidth.

$$f_m = 15 \text{ kHz}; f_d = 75 \text{ kHz}$$

$$m_f = \frac{f_d}{f_m} = \frac{75}{15}$$

hence, $m_f = 5$ **Ans**

$$B = 2(f_d + f_m) \text{ hertz} = 2(75 + 15) \text{ kHz}$$

so $B = 180$ kHz **Ans**

Note: The answer to the above example may be confirmed from Fig. 10.10, as follows:

At $m_f = 5$, it may be seen that the significant modulation components are the carrier, the first, third, fourth, fifth and sixth side frequencies. Thus the highest frequency component to be transmitted is six times the modulating frequency.

Hence, the frequency range required is

$$f_c \pm 6f_m \text{ hertz}$$
$$= f_c \pm 90 \text{ kHz}$$

which gives a bandwidth of 180 kHz.

10.7 Power in the fm Waveform

Since the amplitude of the fm wave remains constant, equal to the amplitude of the carrier, then the power contained in the waveform remains unchanged, regardless of the modulation index. Thus the power is determined only by the rms value of the carrier and the circuit resistance.

10.8 Comparison of Amplitude and Frequency Modulation

Generally, the bandwidth requirement for an fm system is greater than that for am. In the case of vhf broadcasting there is a large difference between the two systems, whereas for nbfm, there is hardly any difference in bandwidth. The difference in bandwidth for broadcasting is illustrated by worked example 10.5. If am was used, the bandwidth would be $2f_m = 30$ kHz; whereas the fm system requires 180 kHz.

An fm wave is far less prone to interference due to noise than is an am wave. The reason is that noise tends to affect the amplitude of the waveform. The amplitude of the frequency-modulated wave is relatively unimportant; whereas it is vital in am. The affect of noise on a signal is usually referred to as the signal-to-noise ratio; so an fm system has a better signal-to-noise ratio than has an am system.

Considering the power contained in the two different types of waveform, it has been shown that for am the power depends upon the modulation index as well as the rms value of the carrier voltage. In the case of fm the power is constant. For this reason fm transmitters are more efficient than am transmitters.

10.9 Phase Modulation

In this form of modulation the instantaneous phase of the carrier varies according to the amplitude of the modulating signal, the number of phase changes per second depending upon the frequency of the signal. In its simplest form it is used for transmitting digital data, so that each time the data change from a binary 1 to a binary 0 (or vice versa) the phase is changed by 180°. Phase modulation is also known as phase shift keying, and is often referred to by the letters 'psk'. The principle of psk is illustrated in Fig. 10.11. In order to successfully demodulate a psk waveform, the phase of a reference waveform needs to be common to both the transmitter and receiver. This presents a considerable problem, and to overcome this a system known as differential phase shift modulation (dpsk) is used. With this system it is the *changes* of phase that represent the two logic levels, and not the phase relative to some reference.

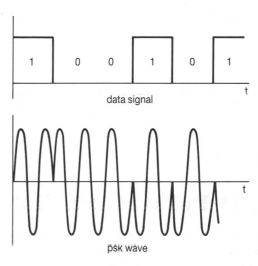

Fig 10.11

10.10 Frequency-shift Keying (fsk)

This is a special form of frequency modulation, where the carrier frequency is deviated to either one of two frequencies, on either side of the carrier frequency. One of these two frequencies represents binary 0 and the other binary 1. The upper and lower frequencies used, and hence the bandwidth required, depends upon the number of binary digits (bits) to be transmitted per second i.e. the bit-rate. The principle of fsk is shown in Fig. 10.12. This form of modulation is used to transmit low-speed data over the public switched telephone network (PSTN).

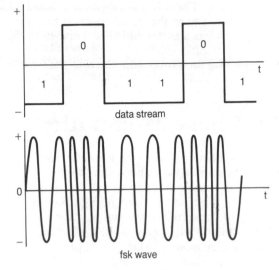

Fig 10.12

10.11 Pulse Modulation

There are four main forms of pulse modulation, although two are really a combination of the others. Essentially, a pulse has two major characteristics: its amplitude and its duration. Thus we have pulse amplitude modulation, pulse duration modulation, and from these we also have pulse position modulation and pulse code modulation. The last of these is the most widely used.

■ Pulse amplitude modulation (pam) ■

Figure 10.13 illustrates the principle of pam, whereby the amplitude of a regular pulse train varies according to the instantaneous amplitude of the modulating signal. When the signal has a positive value the pulse amplitude is increased, and when negative the pulse amplitude is decreased proportionately. The pam waveform will contain the following frequency components:

(i) a d.c. (0 Hz) component of value equal to the mean or average value of the waveform.

(ii) the modulating frequency.

(iii) the pulse repetition frequency (prf), which is the number of pulses per second.

(iv) am sidebands, centred on the prf.

(v) am sidebands centred on integer multiples of the prf.

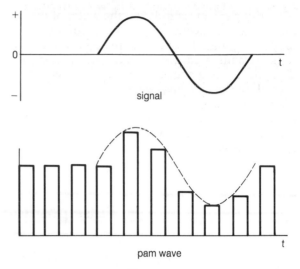

signal

pam wave

Fig 10.13

These different frequency components are best illustrated in a frequency spectrum diagram as in Fig. 10.14. In this diagram, f_r represents the prf, and f_1 and f_2 represent the lowest and highest frequencies in the modulating signal. Provided that there is a space between the lowest frequency in the lower sideband of the prf $(f_r - f_2)$ and highest frequency in the modulating signal, f_2, then the wave can be demodulated by passing it through a low-pass filter. This spacing is assured if $f_r \geqslant 2f_2$.

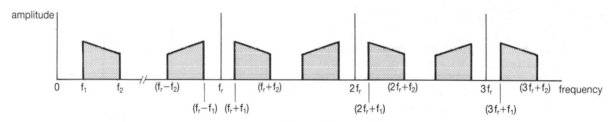

Fig 10.14

■ Pulse duration modulation (pdm) ■

With this form of modulation the width of the pulses is controlled by the amplitude of the modulating signal. To enable the pdm waveform to be demodulated, the prf must be kept constant. Thus the leading edges of the pulses are equally spaced, but the trailing edges will depend upon the individual pulse widths. The pdm wave contains frequency components which include the signal frequency, the prf, sidebands, etc. The demodulation process may again be achieved by

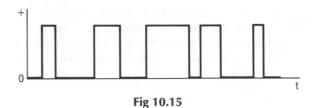

Fig 10.15

passing the waveform through a low-pass filter. A typical pdm waveform is shown in Fig. 10.15.

■ Pulse position modulation (ppm) ■

This is an extension of pulse duration modulation, whereby the pdm waveform is passed through a differentiating network. This results in a series of positive and negative-going voltage 'spikes', as illustrated in Fig. 10.16(a). The positive spikes will be equally spaced since they coincide with the leading edges of the pdm wave. These spikes therefore occur at the prf. The negative spikes coincide with the trailing edges of the pdm wave, and therefore depend upon the modulating signal. Since the positive spikes are not required, the pdm wave is passed through a rectifier. The output of the rectifier is the ppm waveform, consisting of the negative-going spikes, as shown in Fig. 10.16(b). The demodulation process involves re-creating the pdm wave, and passing this through a low-pass filter. The re-creation of the pdm waveform is necessary because the ppm wave does not contain the actual modulating signal frequency. An alternative method of demodulation is to use a low-pass filter followed by an fm demodulator.

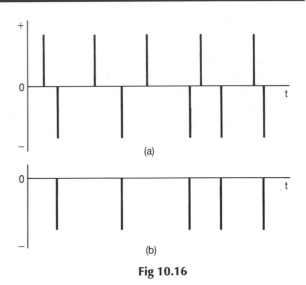

(a)

(b)

Fig 10.16

■ Pulse code modulation (pcm) ■

This is a widely used form of pulse modulation that is an adaptation of pam. The amplitude of the signal to be transmitted is firstly converted into a pam wave. This pam signal is then sampled at regular intervals, and each sample is converted into a series of binary-coded pulses. The resulting pulse train is then transmitted. In order to produce the code, the permissible range of signal voltages is divided into a number of *quantum* levels. Each level is represented by a unique binary code, the most significant bit of which indicates the polarity of the signal. For example, if the msb is a binary 1, then the signal is positive; and if the msb is binary 0, then a negative polarity applies. The larger the number of quantum levels the more accurately can the signal be translated. The number of levels is equal to $(2^n - 1)$, where n is the number of bits in the codes used to represent the levels. This is illustrated in Fig. 10.17. Thus, if there were 8 bits, there would be a total of 255 quantum levels. In Fig. 10.17 only 4 bits are used, hence the total of 15 quantum levels shown.

It may be seen that each quantum level represents a unique voltage level. However the signal voltage amplitude, at the instant of sampling, is quite likely to be at a value somewhere between two quantum levels. In this instance, the amplitude of the pam pulse is *quantised* (rounded off) to the nearest quantum level value. Fig. 10.18(a) shows a signal superimposed on the graph of quantum levels, which is to be sampled at regular time intervals t_1, t_2, etc. The corresponding quantised voltages and binary codes are shown in Table 10.1, and the first six binary 'words' transmitted are shown in Fig. 10.18(b). In this last diagram a binary 1 is represented by a positive pulse, and a binary 0 by a negative pulse.

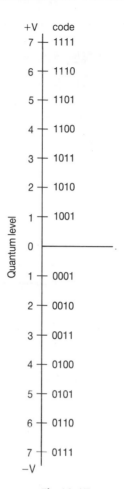

Fig 10.17

Table 10.1   Levels and codes

Time	t_1	t_2	t_3	t_4	t_5	t_6	t_7	t_8	t_9
Quantum level	+2	+4	+6	+4	+2	−1	−4	−3	+1
Binary code	1010	1100	1110	1100	1010	0001	0100	0011	1001

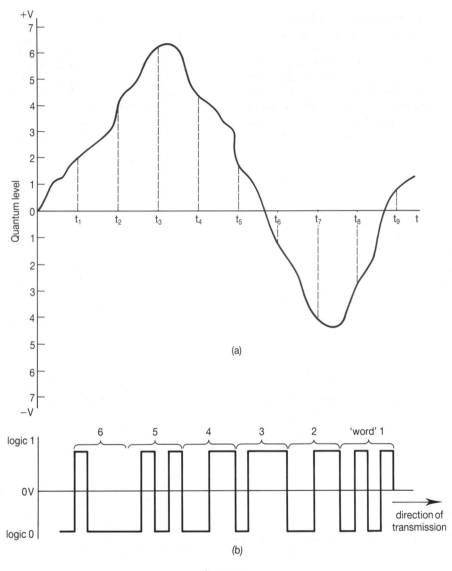

Fig 10.18

10.11 Time-division Multiplexing

Pulse code modulation is used in the PSTN, where the signals are transmitted within the voice frequency (vf) spectrum up to 3.4 kHz. The minimum sampling frequency for a pcm system should be at least twice the highest frequency in the modulating signal, so it needs to be ≥6.8 kHz. In practice, a sampling frequency of 8 kHz is used.

Since the reciprocal of frequency is time, then this means that the signal will be sampled every 125 μs. Since there are 256 quantisation levels, then there are 8 bits in the binary code for each level. Each of these bits occupies a time slot of 488 ns, so each code 'word' occupies a time slot of 3.9 μs. This is only a small proportion of the

125 μs sampling time available, and there are a further 31 time slots (of 3.9 μs each) being unused. Thus, a further 31 channels can be accommodated, resulting in a 32-channel tdm transmission circuit. In practice only 30 channels are used to carry the pcm signals, the remaining two being reserved for synchronising purposes.

At the receiving end a decoder converts the pcm signals into pam signals, which are then routed via a gating system to the individual channel demodulators. As each channel is sampled at 8000 Hz and each one uses an 8-bit code, then the *bit-rate* per channel is 64 kb/s. The transmission lines utilise pulse regenerators at regular intervals to counteract the attenuation and distortion of the digital signal. This also helps to minimise noise effects. Digital data may be transmitted directly along the 32-channel tdm system, without the need for the channelling equipment.

Assignment Questions

1 An am waveform is represented by the expression:

$$v = (10 + 4 \sin 1.6 \times 10^3 \pi t) \sin 10^7 \pi t \text{ volt}$$

Determine (a) the amplitudes of the signal and the unmodulated carrier, (b) the frequencies of the signal and the carrier, (c) the modulation index, and (d) the peak amplitude of the am waveform.

2 A 2 MHz carrier is amplitude modulated, in turn, by (i) a 15 kHz sinewave, and (ii) a complex waveform of fundamental frequency 5 kHz and containing up to the fourth harmonic. Determine, for each case, (a) the highest frequency component in the am wave, and (b) the bandwidth required.

3 The amplitude of an am waveform varies between 16 V and 6 V. Calculate (a) the amplitudes of the modulating signal and unmodulated carrier, and (b) the modulation index.

4 State the condition which produces overmodulation in an am system, and explain why it is an undesirable effect.

5 An am wave is developed across a 200 Ω resistor. If the unmodulated carrier has an amplitude of 12 V, and the signal amplitude is 9 V, determine (a) the modulated index, (b) the power dissipated by the modulated waveform, and (c) the power dissipation when the modulating signal is switched off.

6 With the aid of sketched waveforms explain the difference between amplitude, frequency and phase modulation.

7 A sinewave of amplitude 2.8 V is applied to an fm modulator of sensitivity 20 kHz/V. If the modulation index is 8, calculate (a) the frequency deviation, (b) the signal frequency, and (c) the bandwidth of the fm wave.

8 An fm system has a deviation ratio of 6.5 and a rated deviation of 75 kHz. Determine the maximum allowable modulating frequency component.

9 A complex waveform comprises a fundamental frequency of 2.5 kHz, plus the second, fourth and fifth harmonics. When this waveform is used to frequency modulate a carrier the resulting bandwidth is 165 kHz. Calculate the maximum frequency deviation.

10 Compare amplitude and frequency modulation in terms of the bandwidth requirements, the signal-to-noise ratio, and the efficiency of the system.

11 Explain the difference between pulse amplitude, pulse duration and pulse position modulation techniques.

12 A pcm system utilises 63 quantum levels. Explain what this means, and determine the number of binary digits (bits) used in the coding process.

▪ Answers to Assignment Questions ▪

▪ Chapter 1 ▪

[1] (a) 1.2 A; 1.6 A (b) 2 A (c) 60 Ω **[2]** (a) 2.5 A; 3.183 A (b) 4.05 A (c) 12.35 Ω (d) 125 W **[3]** (a) 5 A; 8.66 A (b) 91.89 mH (c) 0.5; 60° **[4]** (a) 0.667 A; 0.942 A (b) 1.54 A (c) 8.66 Ω (d) 6.67 W (e) 13.85 VA; 12.14 VAr **[5]** 7.96 μF; 100 Ω **[6]** (a) 0.531 A; 0.628 A (b) 0.097 A (c) 513.8 Ω (d) 0 W

[7] (a) 2.23 A (b) 1.18 A (c) 1.284 A (d) −39.45°; 194.7 Ω (e) 248 W (f) 321.2 VA; 301.4 VAr

[8] (a) 15 μF (b) 1.19 A (c) 0.457 A (d) −21.67° (e) 84.9 W **[9]** (a) 18.96 kHz (b) 59.6 (c) 11.9 V (d) 6.67 mA **[10]** (a) 2.03 μF (b) 0.4 A (c) 62.84 V (d) 63.13 V **[11]** (a) 100 Ω; 22.5 mH (b) 2.12 (c) 707 Hz **[12]** (a) 711 Hz (b) 595.5 Hz (c) 17.9; 1.83 **[13]** (a) 1.46 kHz (b) 25 kΩ (c) 0.48 A (d) 7.91 **[14]** (a) 0.833 μF (b) 1.11 mA (c) 60 (d) 33.3 Hz **[15]** (a) 34.72 A (b) 368.4 μF (c) 20.83 A **[16]** (a) 234.6 μF (b) 4.25 kVAr (c) 23.14 A **[17]** (a) 24.45 A; 0.9425 (b) 5.33 μF

▪ Chapter 2 ▪

[1] 3.1 A **[2]** 5.37 A; 9.3 A **[3]** (b) 520 V **[4]** (a) 70.7 A (b) 122.5 A (c) 52.5 kW **[5]** (a) 40.9 A (b) 40.9 A (c) 17.55 kW **[6]** 11.72 Ω; 0.1 H **[7]** (a) 51.15 A (b) 0.866 (c) 31.4 kW **[8]** (a) 90.56 A (b) 52.3 A **[9]** (a) 346.4 V (b) 7.34 A (c) 4.24 A (d) 40.44 kW; 0.53 **[10]** (a) 135 Ω; 0.207 H (b) 3.11 kW **[11]** (a) 1.11 kW (b) 0.736 **[12]** (a) 8.2 kW (b) −70°; 0.342 **[13]** (a) 180 W (b) 0.866 **[14]** 2.17 kW; 13.44 kW **[15]** (a) 19.5 A (b) 30.14 kW

▪ Chapter 3 ▪

[1] 1.314 A **[2]** 0.116 A from B to A **[3]** 5.11 V, A positive **[4]** 0.635 A **[5]** 1.314 A **[6]** 0.116 A from B to A **[7]** 1.2 V **[8]** (a) E_o = 13.24 V; R_o = 4.71 Ω (b) I_{sc} = 1.25 A; R_o = 4.71 Ω (c) 4.71 Ω **[9]** 8.45 V; 2.17 A **[10]** (a) 5.19 V (B positive); 1.037 A (B to A) (b) 0.4 Ω (c) 16.8 W **[11]** 2:1 **[12]** (a) 14 dB

(b) −4.4 dB (c) 24.1 dB (d) 60 dB (e) −6.6 dB (f) 19.3 dB **[13]** 1.052 V **[14]** 0.28 V **[15]** (a) 20 dBm
(b) 4 dBm (c) −8.2 dBm (d) −6 dBm (e) 40 dBm **[16]** 5.13 mW **[17]** +15 dB **[18]** (a) 10 V
(b) −9.54 dB **[19]** −7.5 dB **[20]** −11.1 dB; 0.556 V

■ Chapter 4 ■

[1] (a) 0.183 s (b) 6.15 mA; 0 A (c) 0.915 s **[2]** (a) 333.3 A/s (b) 1 A (c) 15 ms **[3]** (a) 10.1 mA (b) 3.25 V
(c) 22.81 V **[4]** 73 V; 1.03 mA **[5]** (a) 77.1 V (b) 15.45 s **[6]** (a) 2.5 Ω (b) 9 A **[7]** (a) 100 Ω (b) 1 mA
(c) 9.49 s (d) 90.7 μA **[8]** (a) 4 min (b) 1.8 h **[9]** 0.125 A **[10]** 123.32 kΩ **[11]** (a) 416.7 A/s (b) 5.5 ms
(c) 1.216 A **[12]** (a) 0 (b) 110 V; (c) 48.35 V **[13]** (a) 30 H (b) 0.547 s (c) 3.02 A **[14]** (b) 0.881 C
(c) 6.95 V **[16]** 100 Ω

■ Chapter 5 ■

[1] 75 **[2]** 75 **[3]** 0.792 A **[4]** (a) 1800 V (b) 33.75 A (c) 60.75 kW **[5]** 1.2 mWb **[6]** 1.2 kV; 4 kV
[7] (a) 0.182 (b) 97.1% (c) 480 W **[8]** (a) 95.24% (b) 149.4 W (c) 95.62% **[9]** (a) 206 (b) 1500 rev/min
[10] (a) 6 poles (b) 9.38 mWb **[11]** (a) 1332 V (b) 2307 V (c) 31.15 kW **[12]** (a) 14.525 kW (b) 1.825 kW

■ Chapter 6 ■

[1] 16.7 mWb **[2]** (a) 1020 rev/min (b) 340 rev/min **[3]** 415.6 V **[4]** (a) 0.2 Ω (b) 500 W **[5]** (a) 280.75 V
(b) 83.3% **[6]** (a) 5 kW (b) 280 V **[7]** 632 rev/min **[8]** 532 rev/min **[9]** (a) 14.57 mWb (b) 92.74 Nm
[10] (a) 11.79 kW (b) 84.86% (c) 42.82 A **[11]** (a) 4.08 kW (b) 85% **[12]** 91.5% **[13]** (a) 54 kW; 90.5%
(b) 154.92 A

■ Chapter 8 ■

[2] 1.38 Hz **[3]** 355.3 N/m **[4]** 0.38 Hz **[6]** (a) 3.33 m/s (b) 2.1 s (c) 1.59 m/s **[7]** (a) 0.318 Hz (b) 0.333
(c) 0.300 Hz (d) 0.052 m (e) 9.6 Ns/m

■ Chapter 10 ■

[1] (a) 4 V; 10 V (b) 8 kHz; 5 MHz (c) 0.4 or 40% (d) 14 V **[2]** (a) (i) 2.015 MHz, (ii) 2.02 MHz
(b) (i) 30 kHz (ii) 40 kHz **[3]** (a) 5 V; 11 V (b) 0.455 **[5]** (a) 0.75 (b) 461.25 mW (c) 360 mW
[7] (a) 56 kHz (b) 7 kHz (c) 126 kHz **[8]** 11.54 kHz **[9]** 70 kHz **[12]** 6 bits

▪ Index ▪

A.C bridge 152
A.C circuits 1–27
  bandwidth 12
  dynamic impedance 15
  parallel 2–9
  power factor correction 17
  Q-factor 11
  R and C in parallel 2
  R and L in parallel 4
  R, L and C in parallel 6
  resonant, parallel 14
  resonant, series 9
  resonant frequency 10, 15
  series (summary) 1
  three-phase 28
ACIA 175
Active filter 20
Accuracy 146
Alternator 28, 95–98
  emf equation 96
  losses 97
  single-phase 28, 95
  three-phase 30, 95
  rotor construction 95
Amplitude modulation (a.m) 184–189
  bandwidth 186
  modulation index 187
  over modulation 188
  power carried 189
  sideband 185
  side frequency 185
Armature 114, 115
  reaction 116
  windings 119
Asynchronous data transmission 176
Attenuator 63
  asymmetrical 66
  balanced 66
  pad 66
  step 66

Back-emf 128
Bandwidth 12, 16
Baud 177
Bel 59
Bus network 180

CCITT 177
Commutator 114
Compound generator 125
Compound motor 131
Complex waveform, measurement 150
Control systems 153–172
  analogies 153
  computer 171
  damped frequency 160
  damping coefficient 154
  damping ratio 160
  damping term 159
  derivative control 168
  differential equation 154
  driven damped 163
  elements 153
  energy storage 153
  first-order 154
  ideal second-order 157
  inertia term 159
  integral control 170
  natural frequency 159
  practical second-order 159
  proportional control 166
  response curves (amplitude) 162
  response curves (phase) 165
  step input 154
  stiffness term 159
  tachogenerator 168
  three-term controller 170
  time constant 154, 155
  velocity lag 169
Constant current source 51
Constant voltage source 50
Cumulative compounding 125, 131
Current generator 51

Damping 154, 160
Damping ratio 160
Data transmission 175
  asynchronous 176
  synchronous 176
dB 59
dBm 63
D.C generator 120–127
  armature 114, 115
  armature reaction 116
  commutation 115

D.C. generator, *continued*
  commutator   114
  compound   125
  construction   118
  efficiency   134
  emf equation   120
  GNA   117
  interpoles   118
  lap winding   119
  MNA   117
  self-excitation   123
  separately excited   121
  series   124
  shunt   123
D.C transients   70–87
  C–R series circuits   70
  L–R series circuits   77
Decibel (dB)   59–63
  power ratios   59
  voltage and current ratios   61
Delta connection   34
Demodulation   189
Derivative control   168
Differential compounding   126, 131
Differential phase shift keying (dpsk)   195
Differentiating circuit   82
Digital signals   174
Digital voltmeter   149
Dynamic impedance (resistance)   15

Electronic voltmeter   149
Errors (in measurements)   142–148
  calibration   142
  observational and random   146
  systematic   144
  total   147

Faceplate starter   133
Filters   20–24
  active   20
  band-pass   22
  band-stop   23
  coupled tuned circuit   24
  high-pass   21
  low-pass   20
First-order control system   154
Frequencies
  in am waveform   185
  in fm waveform   193
  in pam waveform   197
Frequency deviation   190
Frequency modulation (fm)   190–194
  bandwidth   193
  deviation ratio, D 191
  frequencies carried   193
  frequency deviation, f_d 190
  frequency swing   191

modulation index, m_f 191
modulator sensitivity   190
narrow band fm (nbfm)   193
power carried   194
rated maximum modulating frequency,
  f_{mr} 191
rated system deviation, f_{dr} 191
signal-to-noise ratio   195
Frequency response curves   12, 62
Frequency shift keying (fsk)   196
Frequency spectrum   185

Generator – *see* D.C. generator

Half-power points   62
Handshaking   175
HDLC   178

Induction motors   101–103
  cage rotor   102
  power flow   103
  principle of operation   102
  slip   103
  single-phase   107
  starting methods   105
  three-phase   101
  wound rotor   102
Integral control action   170
Integrating circuit   82
ISDN   174

LAN   182
Lap winding   179
Line current   32, 35
Line voltage   32, 34
Lissajous figures   151
Loading effect   145
Local area network (LAN)   182
Long shunt   125, 131

Maximum efficiency   93, 98
Maximum power transfer theorem   57
Measurements   142–152
  a.c bridge   152
  accuracy   146
  calibration errors   142
  complex waveforms   150
  digital voltmeter (DVM)   149
  electronic voltmeter   149
  frequency and phase   150
  Lissajous figures   151
  loading error   145
  observational errors   146
  power in three-phase circuit   36, 39–41
  random errors   146

sensitivity   146
systematic errors   144
total errors   147
wattmeter corrections   148
Mesh network   181
Modulation   184–201
amplitude   184
frequency   190
phase   195
pulse   196
Modulation index   187, 191
Motors   128–132
compound   131
induction   101, 107
separately excited   132
series   129
shaded pole   108
shunt   129
split phase   107
stepper   137
synchronous   104, 107
universal   108

Natural frequency   159
Network theorems   48–58
Maximum power transfer   57
Norton's   55
superposition   48
Thévénin's   52
Network topologies   180–182
Neutral   31
Neutral current   31, 42
No-volt relay (NVR)   133
Norton's theorem   55

Observational errors   146
Operational amplifier   20, 170
Oscilloscope   150
Overload relay (OLR)   133
Over modulation   188

Parallel a.c circuits   2–9
Parallel resonance   14
dynamic impedance (resistance)   15
resonant frequency   15
Phase current   32, 35
Phase measurement   150
Phase modulation   195
Phase sequence   30
Phase voltage   30
PID controller   170
Power factor correction   17
Power measurement, three-phase   39
Proportional control   166
Protocols   177
PSTN   173

Pulse amplitude modulation (pam)   197
Pulse code modulation (pcm)   198
Pulse duration modulation (pdm)   198
Pulse position modulation (ppm)   198

Q-factor   11

Rated system deviation   191
Reactance   1
Resonance
in mechanical systems   165
parallel   14
series   9
Resonant frequency
in mechanical system   165
parallel circuit   15
series circuit   9
Response curves
amplitude   162
band-pass filter   23
band-stop filter   24
generalised   162
high-pass filter   23
low-pass filter   20, 21
phase   165
second-order system   160, 162, 165
Ring network   180
Rotor
cage   102
cylindrical   95
salient pole   95
wound   102
Rotating magnetic fields   98–101

Salient pole rotor   95
Second-order systems   157–165
ideal   157
practical   159
Selectivity   12
Self-excitation   123
Separately-excited generator   121
Separately-excited motor   132
Series generator   124
Series motor   129
Shaded pole motor   108
Short shunt   125
Shunt generator   123
Shunt motor   129
Sideband   185
Side frequencies   185
Slip   103
Speed control of motors   137
Speed equation   128
Split-phase motor   107
Star connection   31
Star network   181

Starter
   d.c faceplate   133
   star-delta   106
Stepper motor   137
Step input   154, 155
Stiffness term   159
Superposition theorem   48
Synchronous data transmission   176
Synchronous motor   104
Synchronous speed   104
Systematic errors   144
Systems   153–165
   equation   154
   first-order   154
   ideal second-order   157
   practical second-order   159

Tachogenerator   168
Thévénins's theorem   52
Three-phase circuits   28–47
   delta connection   34
   line current   32, 35
   line voltage   32, 34
   measurement of power   39
   neutral   31
   phase current   32, 35
   phase voltage   30, 34
   power dissipation   36
   power factor measurement   41
   star connection   31
Three-term controller   170
Time constant   72, 75, 154, 155
Time division multiplexing   174
Torque equation   128
Transformers   88–94

copper losses   91
efficiency   93
emf equation   88
iron losses   90
measurement of losses   91
on load   91
on no-load   90
matching   57, 90
open-circuit test   92
short-circuit test   92
Transient response   71
Transmission systems   173–183
   asynchronous   176
   parallel   175
   serial   175
   synchronous   176
Tree network   182
Two-wattmeter method   39

UART   175
Universal motor   108
Upper sideband   185

Velocity lag   169
Voltage source   50
Voltage magnification   11
Voltmeter
   digital (DVM)   149
   electronic   149
   frequency response   149

WAN   182
Wattmeter corrections   148
Wave winding   119
Wide area network   182